Essentials of Creatine in Sports and Health

STRENGTHPRO®

StrengthPro® is a proud sponsor of the International Society of Sports Nutrition. This textbook was made possible via an educational grant from StrengthPro®.

Contact information for StrengthPro®
www.strengthpro.com
StrengthPro Inc.
6640 S. Tenaya Way Suite 150
Las Vegas, NV 89113
info@strengthpro.com

Essentials of Creatine in Sports and Health

Edited by

Jeffrey R. Stout, PhD
Department of Health and Sports Science
University of Oklahoma, Norman, Oklahoma

Jose Antonio, PhD
International Society of Sports Nutrition
Deerfield Beach, Florida

and

Douglas Kalman, PhD, RD
Division of Nutrition and Endocrinology
Miami Research Associates
Miami, Florida

Humana Press

Softcover reprint of the hardcover 1st edition 2008
999 Riverview Drive, Suite 208
Totowa, New Jersey 07512

Cover design by Shirley Karina.
Production Editor: Amy Thau

For additional copies, pricing for bulk purchases, and/or information about other Humana titles, contact Humana at the above address or at any of the following numbers:
Tel.: 973-256-1699; Fax: 973-256-8341; or visit our website at http://humanapress.com

This publication is printed on acid-free paper. ∞
ANSI Z39.48-1984 (American National Standards Institute) Permanence of Paper for Printed Library Materials.

10 9 8 7 6 5 4 3 2 1

ISBN 13: 978-1-61737-719-8 e-ISBN 13: 978-1-59745-573-2

Preface

The Dietary Supplement Industry is a booming $21.3 billion dollar industry in the United States alone. This industry includes vitamins, minerals, herbs, specialty supplements, meal replacement powders and bars, as well as sports nutrition and weight loss supplements. There is perhaps no single ingredient in the history of dietary supplements that has been studies more thoroughly than creatine monohydrate. In a nutshell, creatine works and is safe. A PubMed search of "creatine monohydrate" alone brings up 234 research articles, and there are certainly many more peer-reviewed publications on creatine that do not appear on the NLM and NIH databases.

With all of the misinformation regarding the effects of creatine supplementation on health and sports performance, *Essentials of Creatine in Sports and Health* brings together the information on how creatine affecs body composition, exercise performance, and health. Supported by the International Society of Sports Nutrition, this book is timely and vital for all professionals in the field of sports nutrition. Proper supplementation with creatine can improve performance in both endurance and strength-power sports (see chapters by Drs. Willoughby and Cramer). As far as safety, perhaps no other single supplement has been shown to be extraordinarily safe while still being effective. Dr. Poortmans' chapter edifies the reader on the safety and health data regarding creatine.

This text will bring the student, academic scientist, clinician, and sports nutritionist up to date on the latest science of creatine, as well as dispel the common myths that have pervaded the mainstream press regarding this truly remarkable ergogenic aid.

***Jeffrey R. Stout,* PhD**
***Jose Antonio,* PhD**
***Douglas Kalman,* PhD, RD**

Contents

Contributors

JOSE ANTONIO, PhD • *International Society of Sports Nutrition, Boca Raton, Florida*

JOEL T. CRAMER, PhD • *Department of Health and Exercise Science, University of Oklahoma, Norman, Oklahoma*

KEVIN D. BALLARD, MS • *Department of Kinesiology, University of Connecticut, Storrs, Connecticut*

JOAN ECKERSON, PhD • *Department of Exercise Science and Athletic Training, Creighton University, Omaha, Nebraska*

CASSANDRA E. FORSYTHE, MS • *Department of Kinesiology, University of Connecticut, Storrs, Connecticut*

MARC FRANCAUX, PhD • *Institut Supérieur d'Education Physique et de Réadaption, Faculté de Médecine, Université Catholique de Louvain, Louvain-la-neuve, Belgium*

MIKE GREENWOOD, PhD • *Department of Health, Human Performance, and Recreation, Baylor University, Waco, Texas*

DOUGLAS KALMAN, PhD, RD • *Division of Nutrition and Endocrinology, Miami Research Associates, Miami, Florida*

JACQUES R. POORTMANS, PhD • *Institut Supérieur d'Education Physique et de Kinésithérapie, Université Libre de Bruxelles, Brussels, Belgium*

JEFFREY R. STOUT, PhD • *Department of Health and Sports Science University of Oklahoma, Norman, Oklahoma*

JEFF S. VOLEK, PhD, RD • *Department of Kinesiology, University of Connecticut, Storrs, Connecticut*

JOSEPH P. WEIR, PhD • *Physical Therapy Program, Des Moines University–Osteopathic Medical Center, Des Moines, Iowa*

DARRYN S. WILLOUGHBY, PhD • *Department of Health, Human Performance, and Recreation, Baylor University, Waco, Texas*

1 Overview of Creatine Metabolism

Jeff S. Volek, PhD, RD,
Kevin D. Ballard, MS, and
Cassandra E. Forsythe, MS

1. INTRODUCTION

Creatine (Cr) was first discovered as an organic constituent of meat some time in the early 1800s. Later in the 1800s, Cr was consistently detected in muscle tissue extracted from various mammals. It was noted that foxes killed in a hunt immediately after running, contained significantly more Cr than normal, providing the first indication that muscular contraction results in an accumulation of Cr. Around the same time, a substance called creatinine (Crn) was detected in the urine and later determined to be a breakdown product of Cr. Phosphocreatine (PCr) was first isolated from muscle tissue in 1927 and found to play an important role in the transfer of energy. Around the same time, two researchers who consumed large quantities of Cr noted that a percentage of the Cr ingested could not be accounted for by excretion in the urine *(1)*. This study was one of the first to indicate that "Cr loading" in muscle is possible when large amounts of Cr are consumed. A great deal of research has been done since this early work to further define the importance of Cr in humans, and the impact of Cr supplementation. In this chapter, the basic metabolism and function of Cr in humans will be overviewed. To what extent and what factors influence blood- and muscle-Cr levels in response to Cr supplementation will be discussed. Also some of the proposed mechanisms that account for the ergogenic effects from Cr usage observed in many studies will be explored.

From: *Essentials of Creatine in Sports and Health*
Edited by: J. R. Stout, J. Antonio and D. Kalman © Humana Press Inc., Totowa, NJ

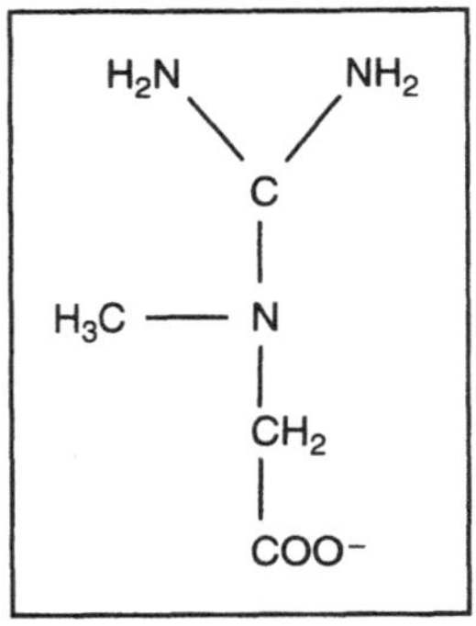

Fig. 1. Structure of Cr.

2. LITERATURE REVIEW

Cr (Fig. 1), also known as methylguanidino acetic acid, is a term derived from the Greek work *kreas*, meaning flesh. Indeed, the highest quantities (~94%) of Cr are in muscular tissues, such as skeletal muscle and heart *(2)*. Cr is an example of a physiologically important nitrogenous compound synthesized from amino acids. However, because rates of Cr biosynthesis are sufficient to maintain normal Cr levels, Cr is not considered an essential dietary nutrient. Cr is also not considered a protein, even though it contains nitrogen and is made from amino acids. Unlike proteins, Cr synthesis does not involve formation of peptide bonds and its degradation does not involve deamination (removal of nitrogen) when it is excreted by the kidney. Thus, the concern that Cr poses a threat to renal function because it represents a nitrogen load on the kidney is unlikely and has not been substantiated (for more information, *see* Chapter 5).

2.1. Endogenous Cr Synthesis

In all mammals, muscle tissues do not have the ability to synthesize Cr, so the compound must be taken up from the blood originating from endogenous biosynthesis or dietary sources. Synthesis of Cr in humans occurs primarily in the liver, from the conditionally essential amino acid arginine, the nonessential amino acid glycine, and the potent methyl donor, S-adenosyl-methionine (SAM) *(2)*. The first step in Cr biosynthesis involves transfer of the amidino group of arginine to glycine, forming L-ornithine and guanidinoacetic acid. This reaction is catalyzed by the enzyme L-arginine:glycine amidinotransferase (AGAT). In the next step, guanidinoacetic acid is methylated by SAM through the action of the

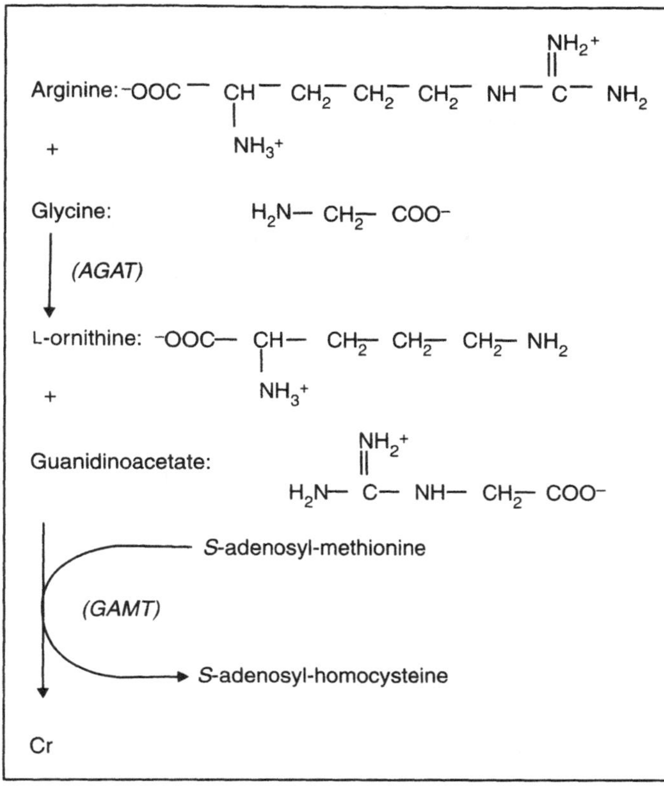

Fig. 2. Biosynthesis of Cr.

enzyme S-adenosyl-L-methionine:N-guanidinoacetate methyltransferase (GAMT) to produce Cr (Fig. 2).

The formation of guanidinoacetate (GAA), through the AGAT reaction, is considered the rate-limiting step in Cr synthesis *(2)*. AGAT is reciprocally repressed by the final end product of the pathway, Cr. This is a classic example of a negative feedback mechanism whereby the product of a reaction inhibits the main enzyme in the biosynthetic pathway. In the case of Cr, this has likely evolved as a way to conserve the dietary essential amino acids, arginine and methionine, during times of dietary insufficiency or stress (including trauma, sepsis, and burns). Through this negative feedback, an increase in the serum concentration of Cr, either because of endogenous synthesis or dietary ingestion, results in a decrease in the mRNA content, enzyme level, and enzymatic activity of AGAT *(3)*. AGAT can also be modulated by hormonal and dietary factors *(3)*.

Growth hormone and thyroid hormone deficiency decrease AGAT activity, whereas conditions of dietary deficiency that increase serum concentrations of Cr (fasting, protein-free diets, and vitamin E deficiency), downregulate AGAT expression *(3)*. Finally, AGAT may be controlled by sex hormones, with estrogens decreasing and testosterone increasing AGAT levels. For example, oral administration of methyl testosterone to healthy humans stimulates AGAT expression and thus, Cr biosynthesis *(4)*. Overall, the rate of Cr biosynthesis is on the order of 1–2 g/d.

In humans, inborn errors in Cr biosynthesis are rare. Two defects have been discovered including one in AGAT identified in 2001 and one in GAMT identified in 1997 *(5)*. In AGAT deficiency, decreased circulating concentrations of Cr and GAA are found. Affected patients show mental and motor retardation and severe delay in speech development. In individuals with GAMT deficiency, there is a wide range of clinical symptoms ranging from developmental delay with absence of active speech, to severe cases of epilepsy. Diagnosis of GAMT deficiency can be made by detection of GAA in plasma or urine *(5)*.

From the liver, Cr can be exported and transported throughout the bloodstream and taken up by Cr-requiring tissues, such as skeletal muscle, heart, and brain *(2)*. Limitations in Cr synthesis might exist during conditions of folic acid and/or vitamin B_{12} deficiency *(6,7)*, and other physiological and pathological conditions in which synthesis of SAM is impaired *(2)*. SAM is necessary to donate a methyl group to the newly formed Cr molecule, and is formed through reactions involving the essential amino acid methionine and the aforementioned B-vitamins. Without SAM, Cr could not be synthesized, which underscores the importance of adequate protein and B-vitamin intake. There are species differences in Cr synthesis, as both the kidney and pancreas have been shown to play a role in production of Cr in various mammals. In this regard, the detailed contribution of different bodily organs to Cr synthesis is still rather unclear *(2)*.

2.2. Dietary/Exogenous Cr

Food products such as red meat and fish provide the human body with an exogenous source of Cr (Table 1). In general, animal flesh products are the highest dietary sources. Red meat for example, contains approx 2–5 g of Cr/kg. Cooking time, temperature, and acid increase the breakdown of Cr to its metabolic byproducts; thus, well-cooked meats have a lower Cr content than rare meats. Average intake of Cr in people who eat meat is about 1 g/d *(8)*. Another way to obtain dietary Cr is through supplements, such as Cr monohydrate.

Table 1
Food Sources of Cr

Food	*Cr content (g/kg)*
Beef	4.5
Cod	3
Cranberries	0.02
Herring	6.5–10
Milk	0.1
Pork	5
Salmon	4.5
Shrimp	Trace
Tuna	4

2.3. Absorption, Transport, and Uptake

Dietary Cr has a very high bioavailability, passing through the digestive tract intact for transport directly into the bloodstream. Normal plasma levels of Cr are about 50 mmol/L, and increase sharply after supplementation with Cr or ingestion of meat. Research with food and supplement sources of Cr has shown that consuming 2 g of Cr in solution resulted in a peak plasma concentration of nearly 400 mmol/L at 30–60 min, whereas consuming a similar amount of Cr from steak resulted in a slightly lower peak plasma concentration that was elevated for a longer period of time following ingestion *(9)*. Supplementation with a 5 g dose of Cr results in peak plasma values approaching 1 mmol/L, and higher doses result in even greater plasma concentrations of Cr. Once absorbed into the bloodstream, Cr (from both endogenous and dietary sources) is either cleared by the kidneys or taken up at the tissue level, primarily skeletal muscle, to be used.

With respect to dietary Cr, not all of the ingested Cr can be retained in the body once absorbed, especially when high doses of supplements are taken. A high proportion of Cr is usually retained in the initial days of Cr supplementation, but urinary Cr excretion progressively increases with continued ingestion *(10–12)*. An explanation for the decrease in Cr retention with high-dose supplementation may be owing to the observation that, Cr supplementation can result in downregulation of the Cr transporter (CreaT) isoform expression in skeletal muscle *(13)*. In other words, there are less CreaT receptors in muscle after supplementing with Cr. This effect on CreaT receptors is probably one reason why there appears to be a limit or maximal amount of Cr that can be stored in skeletal muscle no matter how much Cr is consumed in the diet or how high blood levels of Cr are elevated. However, this effect on CreaT appears to be reversible once Cr supplementation is discontinued.

Cr is actively transported into tissues against a strong concentration gradient by way of a sodium- and chloride-dependent CreaT protein that spans the tissue membrane *(2,5)*, thereby keeping plasma levels relatively low compared with that inside the muscle. It has been shown that more than 90% of the cellular uptake of Cr occurs through these CreaT receptor proteins *(14,15)*. The CreaT receptor is partially regulated by the extracellular Cr concentration in a typical negative feedback manner. A genetic defect in the CreaT has been found with the gene encoding CreaT located on the X-chromosome *(5)*. Affected male patients show mild-to-severe mental retardation, with absence of speech, whereas affected (heterozygote) females have milder symptoms. Some carriers of the defect are without symptoms, whereas others have learning disabilities to various extents.

2.4. Muscle-Cr Concentrations

Skeletal muscle is the primary tissue in which Cr is stored. It is been estimated that 60% of muscle-Cr content is stored in the form of PCr *(12,16,17)*. The average total-Cr content in human-skeletal muscle is approx 120 mmol/kg dry mass, with a range of approx 90–160 mmol/kg dry mass. Normal variability in muscle-Cr may be resulting from a variety of factors such as meat intake, muscle-fiber type, training status, age, gender, and other unknown factors. Skeletal muscle-fiber type possibly affects the content of total-Cr within muscle, as type II fibers has higher levels of Cr and PCr than type I fibers *(17–19)*.

Vegetarians, because of their low consumption of animal products, represent a unique group who rely exclusively on endogenous Cr production to maintain Cr homeostasis. Although plasma Cr levels may be considerably lower in vegetarians compared with omnivores, muscle-Cr concentrations are usually within the normal range, albeit on the low end of the distribution *(20–22)*. One study examined Cr content in meat eaters after consuming a vegetarian diet for 3 wk and found that total-Cr content was 13% lower compared with that before the diet *(23)*. This suggests that synthesis of Cr in the body is adequate to maintain muscle-Cr levels in individuals on a Cr-free diet (i.e., vegetarians), but it will not maintain levels as high as omnivores. Because individuals with lower muscle-Cr concentrations exhibit the greatest accumulation in response to supplementation, vegetarians have been singled out as being a group that should respond most favorably to Cr supplementation.

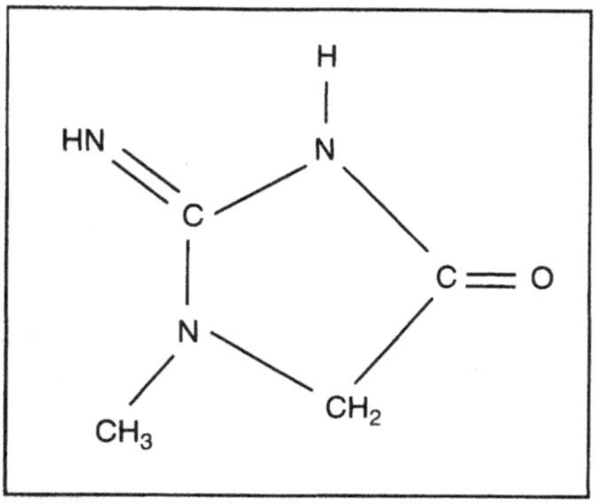

Fig. 3. Creatinine.

2.5. Degradation and Elimination of Cr

The breakdown of Cr and PCr occurs nonenzymatically in the liver to produce the product Crn (Fig. 3). This reaction degrades the muscle-Cr pool at a rate of about 1.6%/d. Assuming that a 70 kg man has approx 120 g of total body Cr, roughly 2 g/d are degraded to Crn and this must be replaced by dietary or endogenous Cr sources *(24)*. As 2 g of Cr is a relatively small amount, and can be replaced easily by the body or diet, muscle-Cr concentrations remain relatively stable and are not significantly influenced by intense exercise or other stressors. Normal muscle-Cr homeostasis is therefore maintained by a balance between dietary intake, biosynthesis, and degradation (Fig. 4).

Once Cr is degraded to Crn, it constantly diffuses out of sites of synthesis into the blood and is excreted in the urine by the kidneys. The reason that Crn is not retained in the cell but is instead lost in the urine, has to do with several factors, including the fact that Crn is membrane permeable *(2)*. Twenty-four hour urinary Crn excretion is frequently used as a rough measure of total muscle mass because the rate of Crn formation from Cr is nearly constant each day, and because more than 90% of total body Cr is found within muscle tissue *(25)*.

2.6. Cr Kinase Reaction

Once inside the cell, a portion of the total body Cr is phosphorylated (i.e., has a phosphate group attached) and is referred to as PCr or Cr phosphate, and is stored within tissues. Phosphorylation of Cr is catalyzed by the enzyme Cr kinase (CK), and is known as the CK reaction (Fig. 5). Several distinct isoforms of CK are known to exist, depending on the species of animal and its developmental stage, and all catalyze

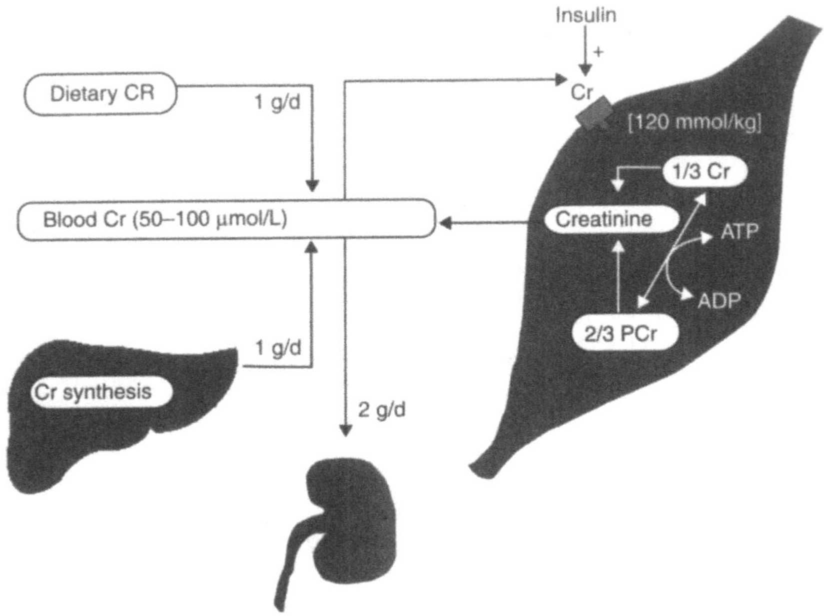

Fig. 4. Normal dietary intake and endogenous biosynthesis of Cr contribute to a muscle uptake of about 2 g Cr/d, which is equal to the amount degraded to Cr and eliminated in the urine.

Rx no. 1 (exercise):	ATP ---*(ATPase)*---> ADP + Pi
Rx no. 2 (exercise):	PCr + ADP + H + ---*(CK)*--->Cr + ATP
Rx no. 3 (recovery):	PCr + ADP + H + <---*(CK)*---Cr + ATP

Fig. 5. CK reaction during exercise and recovery.

the reversible transfer of the γ-phosphate group of ATP to the guanidino group of Cr to yield PCr and ADP *(2)*. During intense exercise, the rate of ATP hydrolysis in the muscle fiber is extremely high, driving the ATP reaction (Rx no. 1) to the right. Consequently, during intense exercise, ADP and inorganic phosphate (Pi) accumulate within cells, but must be removed. To prevent overaccumulation of ADP and Pi and prevent depletion of ATP, PCr serves as an energy buffer through the CK reaction (Rx no. 2). This results in preservation of ATP levels at the expense of depletion in PCr and a stociometric increase in free-Cr.

During recovery from exercise, the CK reaction is reversed and PCr is replenished (Rx no. 3).

PCr resynthesis is rapid with a half-life of about 30 s and about 95% of PCr is resynthesized after only 3–4 min. The advantage of using PCr as an energy buffer in the CK reaction is that PCr is a very high-power energy system (i.e., it produces a large amount of ATP per unit time). However, the storage capacity of PCr is relatively small, meaning that PCr is depleted very quickly during maximal activity (i.e., 10–20 s). Thus, optimizing Cr stores to enhance PCr stores and delay depletion will likely enhance exercise performance.

2.7. Role of Cr and Cr Phosphate

As discussed earlier, Cr, in the form of PCr, is essential for replenishing ATP stores that are used immediately during high-intensity exercise. When PCr is readily available for ATP regeneration, glycolysis is induced with a delay of a few seconds, and stimulation of mitochondrial oxidative phosphorylation is delayed even further. Despite the almost instantaneous availability of PCr for energy production, the PCr stores in muscle are limited, and during high-intensity exercise, PCr is depleted very quickly. However, without this limited supply of PCr, ATP concentrations would diminish rapidly and resynthesis of ATP would be greatly delayed. Improving the status of skeletal muscle-PCr stores through Cr ingestion can help delay PCr depletion *(17)*, and rapidly refresh stores of ATP to prevent the occurrence of fatigue during short-term muscular effort *(26)*. Despite simply improving ATP stores, other roles for increased PCr availability include:

- Diminished reliance on anaerobic glycolysis for energy production and reduction in the associated lactate formation *(27)*.
- Facilitation of muscle relaxation and recovery during repeated bouts of intense, short duration effort through increased rate of ATP and PCr resynthesis, thus allowing for continued high power outputs *(28)*.

Another role of PCr is to serve as an intramuscular high-energy phosphate shuttle between the mitochondria (site of production) and myosin cross-bridge sites that initiate muscular contraction (site of utilization) *(29)*. This mechanism has been termed the PCr energy shuttle, and was first proposed by Bessman in 1954 *(30)*. Muscular contraction occurs when two sets of muscle filaments slide past each other: thick filaments contain the protein myosin and the high-energy molecule ATP, and thin filaments primarily contain the protein actin. The relative sliding of thick and thin filaments is brought about by "cross bridges," parts of the myosin molecules which stick out from the myosin filaments and interact cyclically with the thin filaments, transporting them by a kind of "rowing" action. The energy required for this process to occur is provided by the hydrolysis of ATP catalyzed by myosin ATPase, producing force, ADP, and Pi.

To prevent accumulation of ADP which can potentially reduce muscle contractility, it is rephosphorylated to ATP through the CK reaction, using PCr as a substrate *(31)*. Specific CK isoenzymes are located at the peripheral terminus of the myosin heads, which rephosphorylate the ADP produced during the cross-bridge cycle. Free-Cr is then liberated and diffuses into the intervening space of the muscle fiber, traveling in the opposite direction as PCr, where it finally arrives at the energy-generating terminus of the mitochondria. In the mitochondria, the free-Cr interacts with CK, and PCr is formed from mitochondrial ATP. The PCr is then shuttled back to the sites of utilization (i.e., the myosin head) and the process continues *(32)*. Thus, the PCr energy shuttle connects sites of energy production with sites of utilization by carrying energy through PCr.

2.8. Cr Supplementation: Effects on Muscle-Cr

Evidence that the intramuscular stores of Cr could be increased by ingesting Cr in greater than normal amounts dates back to at least 1926 *(1)*. Several studies had noted that when dietary Cr was increased, there was retention of a significant portion of the ingested Cr in the body, as calculated by excretion of Cr in the urine. The standard method to measure muscle-Cr levels is to obtain a small amount of muscle tissue with a biopsy needle and analyze the Cr content using standard enzymatic assays. Using this technique, several more recent studies have confirmed that supplementation with Cr over a period of several days results in accumulation of Cr in skeletal muscle. Increasing muscle-Cr stores is a major aim of the supplementation regimen, and the extent of muscle-Cr accumulation is related to the magnitude of improvement in exercise performance.

The most common methods used to increase muscle-Cr involves taking multiple 5 g doses every 3–4 h/d for 3–7 d. These doses of Cr are much higher than the amount obtained in a normal diet and therefore, this method is often called "Cr loading." Using this regimen, the majority of Cr accumulation occurs during the first few days of loading, and tapers off so that almost all of the ingested dose can be recovered in the urine after about 5–7 d. Because the majority of Cr in the human body is stored within skeletal muscle, arguments have been made that Cr should be consumed relative to body weight or fat-free mass. A common recommendation is to ingest 0.3 g Cr/kg body wt. This dose was derived from studies that measured muscle-Cr stores in subjects weighing

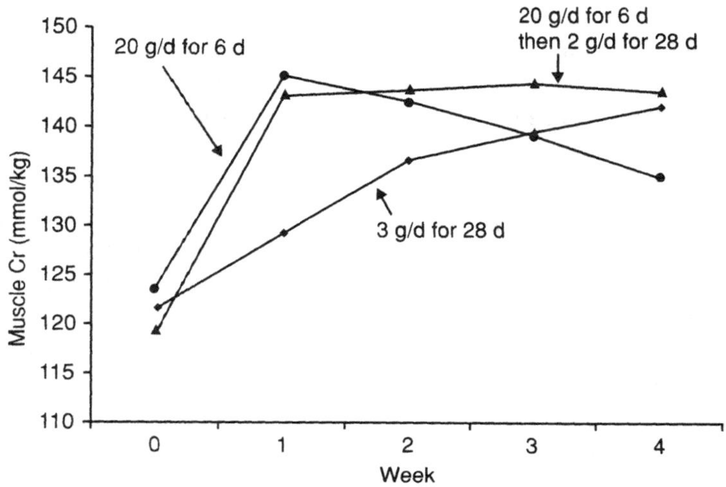

Fig. 6. Muscle-Cr levels in subjects supplemented with varying doses of Cr over a 4-wk period. Data from ref. *21*.

approx 80 kg and supplemented with a dose of 20 g/d (20 g Cr/d divided by 80 kg = 0.3 g Cr/kg). Although valid in theory, studies have not been performed to confirm whether 0.3 g Cr/kg is effective in individuals with a body weight significantly less than or greater than 80 kg.

A loading regimen may not be necessary to increase muscle-Cr levels even though "Cr loading" is a popular dosing method. Ingestion of 3 g of Cr/d for 28 d results in a similar increase in muscle-Cr concentrations (~20%) compared with that obtained during a standard loading protocol followed by a lower maintenance dose *(21)*. Thus, muscle-Cr stores can be increased rapidly by consuming 20 g/d for 6 d or slowly by consuming 3 g/d for 4 wk (Fig. 6). Consuming less than 3 g/d may not result in increased muscle-Cr uptake. For example, ingestion of 3 g of Cr/d for 6 wk failed to alter muscle PCr levels at rest, during exercise, and recovery from exercise in athletic women *(33)*. However, some studies have shown positive performance effects because of small doses of Cr. If supplementation is discontinued after the loading phase, the elevation in muscle-Cr decreases at a slow rate. Even 4 wk after a Cr-loading period, muscle-Cr remains slightly more than presupplementation levels. However, by 4–6 wk any accumulation in muscle-Cr has been washed out. Therefore, a low "maintenance" dose is necessary to maintain elevated muscle-Cr levels.

There remains some uncertainty about the appropriate Cr dose necessary to retain elevated muscle-Cr stores after a loading phase.

It appears 2–3 g/d, the amount normally degraded to Crn, might be sufficient to maintain increased muscle-Cr stores. In one study, a daily 5 g Cr dose following an initial loading dose was sufficient to maintain elevated muscle-PCr stores in untrained women who participated in a 10-wk resistance training program *(10)*. A 5 g maintenance dose following an initial loading dose resulted in a small decline in muscle-Cr in moderately trained men during 12 wk of intense resistance training *(34)*. As was the case for loading doses of Cr, a maintenance dose has also been recommended relative to body weight. Because 2 g/d was adequate to maintain elevated muscle-Cr stores in individuals weighing approx 80 kg, a maintenance dose of 0.03 g Cr/kg body wt might be appropriate (2 g Cr/d divided by 80 kg = 0.03 g Cr/kg/d). Once again, this maintenance dose has not been validated in athletes of different body sizes performing different training programs. Other factors might affect the maintenance dose such as variation in diet composition and habitual Cr intake, muscle fiber type distribution, gender, age, and initial total muscle-Cr concentrations. There is a great deal of variability in the magnitude of increase in muscle-Cr stores associated with Cr supplementation between individuals. Individuals with lower levels of total-Cr generally experience larger increases in muscle-Cr stores compared with individuals with higher starting concentrations *(17,27)*. The increase in muscle-Cr stores is also positively associated with improvements in exercise performance *(17,27)*. Several other factors have been shown to influence the extent to which muscle-Cr stores respond to supplementation.

2.9. Influences on Cr Uptake

2.9.1. Carbohydrate and Insulin

Studies have addressed the influence of insulin on muscle-Cr uptake in humans by measuring muscle- and urine-Cr during Cr supplementation with infused insulin, carbohydrate ingestion, or combined protein and carbohydrate ingestion. In two separate studies Green and colleagues demonstrated reduced urine-Cr losses and increased muscle-Cr accumulation in subjects ingesting Cr concurrently with high dose of carbohydrate (90 g Cr, four times per day) *(20,35)*. More recent studies have examined the effects of carbohydrate/protein combinations *(36)* or carbohydrate alone *(37)* on muscle-Cr uptake. For instance, Steenge et al. *(36)* reported that the ingestion of Cr with 50 g of protein and 50 g of carbohydrate results in similar muscle-Cr increases as ingesting Cr with 100 g of carbohydrate. Preen et al. *(37)* established that Cr supplementation combined with 1 g glucose/kg body mass twice per day increased muscle total-Cr by 9% more than Cr supplementation alone. Thus, Cr accumulation might be increased slightly by ingesting insulin-stimulating nutrients such as carbohydrate.

2.9.2. Exercise

Cr uptake has been shown to be stimulated following submaximal exercise bouts *(12,38)*. This increase in Cr uptake following exercise has been demonstrated to only occur in skeletal muscles involved in the exercise. For example, when subjects performed 1 h of strenuous exercise per day with only one leg, Cr supplementation increased the mean muscle-Cr content from 118 to 149 mmol/kg dry mass in the control leg, but to 162 mmol/kg dry mass in the exercised leg *(12)*. It has been hypothesized that increased Cr uptake with exercise is a result of increased blood flow to the working musculature *(12)*. Other possible explanations for the enhancement of Cr uptake into skeletal muscle with exercise has been suggested to be resulting from activation of CreaT proteins, the synthesis of new transporters, or changes in the forces driving Cr transport *(38)*.

2.9.3. Training

Conflicting reports of changes in skeletal muscle total-Cr concentrations after various types of training (endurance, sprint, and resistance) make it difficult to determine the effects of regular training on Cr muscle uptake. One study found that sports students demonstrated an increased muscle PCr:ATP ratio compared with sedentary subjects following Cr supplementation indicating that individuals involved in regular training may demonstrate a greater ability to accumulate Cr stores *(39)*. The findings that acute exercise stimulates Cr uptake by skeletal muscle, as mentioned in the previous section would suggest that regular training would only further enhance Cr uptake, but more research must be done to validate this hypothesis.

2.9.4. Muscle Fiber Type

Type II (fast-twitch) muscle fibers have higher levels of Cr and PCr ($\approx$12%) compared with type I (slow-twitch) fibers *(24)*. However, limited research in humans has been conducted to determine if a difference exists between muscle-fiber types in regards to Cr uptake. Casey et al. *(17)* found that the PCr content in type I and type II muscle fibers of the quadriceps femoris muscle increased to a similar extent ($\approx$15%) following Cr supplementation. This finding indicates that uptake may not differ between fiber types even though initial-Cr content may be unequal.

2.9.5. Gender

Several studies have demonstrated that skeletal muscle total-Cr content is similar between men and women *(40,41)*. These results are consistent in that the magnitude of skeletal muscle-Cr loading following supplementation *(12,41)* and that CreaT-mRNA expression in skeletal muscle is similar between genders *(40)*. Therefore, women and men should benefit to the same extent from dietary-Cr intake.

2.10. Mechanisms of Action

When Cr accumulates in human-skeletal muscle it produces various effects, which may result in the positive improvements that are seen in exercise performance following Cr supplementation. These positive improvements are most likely because of an enlargement of the phosphagen pool in skeletal muscle, which is used to fuel high-power activities. Recently, several other mechanisms have been put forth to possibly explain improvements in exercise performance and are discussed within this section.

2.10.1. Energetic Theory

The most widely accepted theory explaining the beneficial effects of Cr supplementation on muscular performance is the energetic theory: ↑ muscle-Cr → ↑ training intensity → ↑ training stimulus → ↑ physiological adaptations to training (i.e., ↑ muscle fiber hypertrophy, muscle mass, and strength). Thus, increased Cr stores might allow for an athlete to train at a higher intensity, thereby increasing the stimulus placed on the muscle and increasing the physiological adaptations to that exercise bout.

2.10.2. pH Buffer

During periods of low pH, such as when hydrogen ions from lactic acid production and ATP hydrolysis accumulate during intense exercise, the CK reaction favors the regeneration of ATP (*see* Rx no. 1). Thus, rapid increases in hydrogen ions with intense exercise are buffered by PCr and might help to prevent fatigue. This pH buffer is one mechanism in which Cr maintains optimal exercise performance, as it helps prevent acidification of cells and maintains a normal pH *(42)*. A decreased pH has numerous effects on the intra- and extracellular environments of all cells, and has been associated with contributing to the onset of fatigue by various mechanisms.

2.10.3. Membrane Stabilization

Cr might be useful for preventing tissue damage by stabilizing cell membranes. Cr, in the form of PCr, may decrease membrane fluidity as well as the loss of intracellular enzymes like CK from the cytoplasm *(42)*. This suggests that increased levels of PCr in human-skeletal muscle may provide protection to muscle cellular membranes from the stress of strenuous exercise. Rawson et al. *(43)* recently examined whether exercise-induced muscle damage would be reduced by Cr supplementation in young male individuals. The results of this study demonstrated that 5 d of Cr supplementation did not decrease muscle damage or improve recovery from strenuous exercise (i.e., eccentric contractions) compared with a placebo group. The authors of this study suggested that long-term Cr supplementation should be investigated to determine its effects on muscle damage and recovery rates in athletes undergoing strenuous eccentric exercise programs.

2.10.4. Muscle Relaxation Time

Cr might improve exercise performance by decreasing muscle relaxation time following a contraction. van Leemputte et al. *(28)* determined that 5 d of Cr supplementation in healthy men resulted in no significant change in maximal isometric force but the time needed to relax was significantly decreased. These results demonstrate that Cr might decrease muscle relaxation time in some individuals and that the decrease in muscle relaxation time is greater in those individuals who took longer to relax before supplementation.

2.10.5. Glycogen Storage

There is reason to speculate that Cr supplementation might influence muscle-glycogen levels owing to findings demonstrating that Cr increases intracellular water *(44)* and that cellular water affects glycogen levels in rat-skeletal muscle *(45)*. The majority of studies examining human-skeletal muscle glycogen levels following Cr supplementation have shown positive results *(38,46–49)*. The ergogenic effects of increasing muscle glycogen levels with Cr supplementation are unclear but it may suggest that enhanced glycogen levels could possibly result in training-induced improvements given that improved glycogen stores positively affect high-intensity exercise *(50,51)*, including resistance training *(52,53)*.

2.10.6. Improved Lean Body Mass

A common explanation for increased exercise performance with Cr supplementation in combination with strength training is enhanced muscle-fiber size and increased lean body mass (LBM) *(36,54)*. Several human studies have shown that Cr doses of 20 g/d for 4–28 d results in increases in total body mass of 1–2 kg, with these increases resulting from augmented LBM *(42)*. Resistance training in conjunction with Cr supplementation has been found to result in significant increases in body mass and LBM, as well as types I, IIa, and IIb muscle fiber cross-sectional areas (CSA) in healthy, young males *(34)*. In the study by Volek et al. *(34)*, subjects participated in a 12-wk heavy resistance training program with or without Cr supplementation. Those subjects that supplemented with Cr had greater increases in both type I and type II muscle fiber CSA compared with those who only trained *(34)*.

The increases in LBM seen with Cr supplementation have been suggested to occur owing to several different mechanisms. First, Cr supplementation may increase protein synthesis or reduce protein breakdown *(42)*. A study by Parise et al. *(41)* found that acute Cr supplementation did not result in an increase in muscle fractional synthetic rate but that it may result in a decrease in muscle protein catabolism. Second, several studies have suggested that Cr supplementation along with resistance training improves the expression of myogenic regulatory factors that might possibly enhance the expression of skeletal-muscle myosin heavy chain and CK activity *(55–57)*. Third, accumulation of muscle-Cr levels have been reported to increase intracellular water retention, which is believed to promote an anabolic effect through a positive nitrogen balance and protein synthesis. However, owing to the lack of studies that have examined the influence of intracellular water on skeletal-muscle protein synthesis, this hypothesis must be viewed with care.

2.10.7. Satellite Cell Activity

In adult humans, the cells responsible for skeletal-muscle hypertrophy are satellite cells (SC). These cells are so named because instead of being located inside the muscle fiber, they lie along the muscle-fiber membrane under the basal lamina *(58)*. The role of SC is to provide myonuclei to enlarging muscle fibers when muscle growth is stimulated. When SC are activated with exercise, they proliferate and differentiate to assist in the formation of new muscle fibers. In human studies, resistance training has been shown to increase the number of SC and myonuclei in the trained muscle, indicating that SC play an important role during hypertrophy. The importance of SC for adding myonuclei to growing muscle has also become apparent when SC are inactivated, such as they can be in studies with mice and rats; in these states, muscle fibers only have a limited capacity for growth *(59,60)*.

The ability of Cr to augment muscle hypertrophy has been shown to be in part because of its effects on SC. In rat-skeletal muscle, Cr supplementation during increased functional loading and compensatory muscle hypertrophy caused an increase in SC activity as indicated by a higher number of myonuclei compared with muscle fibers of rats without supplementary Cr *(61)*. Furthermore, the effects of Cr monohydrate on SC were demonstrated in vivo *(62)*. When Cr monohydrate was added to myogenic cell cultures, it significantly increased SC differentiation compared with no Cr or other ergogenic agents, such as l-glutamine. In 2006, Olsen et al. *(63)* demonstrated the effects of Cr monohydrate supplementation on number of SC and myonulclei in human-skeletal muscle during 16 wk of strength training. Thirty two males were given 6 g of Cr monohydrate a day and resistance trained three times a week. The supplement was compared with a protein and carbohydrate supplement, a carbohydrate only supplement, and no training or supplement. Muscle biopsies were taken at week 4, 8, and 16 for analysis, and maximal isometric muscle strength (i.e., maximal voluntary contraction [MVC]) of the right quadriceps femoris was measured after the 16-wk training period. Although all training groups increased SC and myonuclei number, Cr monohydrate supplementation significantly augmented the training-induced increase in SC and myonuclei compared with the other supplementary groups throughout the entire training period

Furthermore, Cr significantly increased muscle-fiber CSA compared with the other groups, which was shown to be positively correlated to the increase in number of myonulcei. Cr also resulted in the highest MVC after the training period. The accelerated and superior gains in number of SC and myonuclei with Cr monohydrate supplementation combined with 4 mo of resistance training suggest that Cr monohydrate increases the contribution of SC-derived myonuclei to the muscle fibers, which is expected to increase the rate of muscle hypertrophy. Indeed, increased muscle-fiber area was found, which was accompanied by a corresponding greater increase in maximal isometric muscle strength (i.e., MVC). The muscle hypertrophy effects of Cr seem to be linked to the activity of exercise training, as Cr supplementation without training does not seem to lead to increases in SC activity *(61)* or muscle-fiber area. In conclusion, Cr monohydrate affects the cellular processes involved with muscle-fiber growth in response to exercise. By accelerating the adaptation of SC to resistance training, Cr can result in increased muscle hypertrophy and possibly, increased muscular strength.

3. PRACTICAL APPLICATIONS

The majority of athletes, both recreational and competitive, considering Cr monohydrate supplementation have very limited knowledge of the mechanisms responsible for its positive benefits toward performance. Most information about Cr is obtained through word of mouth, sales associates, or the popular press; and may not be entirely accurate.

Misunderstandings and anecdotes, both positive and negative, of Cr are plentiful. The idea that Cr induces significant water retention, cramping, and may even cause heat intolerance is one common anecdote that has failed to be substantiated in a number of studies. Also, the muscle-building effects of Cr are commonly thought to be a fallacy, and the increase in weight after Cr supplementation is attributed only to increased fluid retention. Because of these common rumors surrounding Cr usage, many individuals may be cautious of supplementing with Cr and might convince others to avoid this supplement all together. However, certain athletes, particularly those who regularly participate in activities or sports involving high-intensity, short-duration movements (e.g., soccer, basketball, and resistance training) would likely benefit greatly from the use of Cr monohydrate as an ergogenic tool.

For a sports nutritionist to provide accurate information of Cr to athletes of all ability levels, and dispel the plethora of myths surrounding Cr supplementation, they must understand first the fundamental basics of Cr metabolism. The information provided in this chapter can provide sports nutritionists with the proper tools to educate athletes who come to them with questions and interests about Cr monohydrate supplementation. Also, with accurate information supported by research provided in this chapter, the athlete would be able to make an informed decision regarding their usage or nonusage of Cr.

4. CONCLUSION

The increased usage of the supplement Cr monohydrate over the past decade has warranted the need for further research exploring the mechanisms of Cr transport, uptake, and storage and how these factors are influenced by diet and exercise. Recent research has supported the safety and efficacy of Cr supplementation for improving exercise performance in various populations. There is also a better understanding of the mechanisms responsible for these positive performance outcomes. It would be fair to state that Cr supplementation represents one of the most studied dietary supplements, and in particular one repeatedly shown to have ergogenic effects. This does not mean that all athletes will benefit significantly from Cr supplementation only that, in controlled studies subjects supplemented with Cr often outperform and experience improved adaptations to resistance training compared with placebo controls.

4.1. Side Bar No. 1

The media and lay population alike are quick to vilify virtually every dietary supplement as ineffective and dangerous. Although this perspective may prove true in the case of many supplements, it does a serious disservice to Cr, which is undoubtedly the most researched supplement in history. Numerous studies have verified Cr's short-term safety. Although the long-term effects of Cr supplementation have not yet been explored, given that athletes have been utilizing the supplement for almost a decade without noteworthy incident, there appears to be little cause for concern in this regard. The fact that Cr is a naturally occurring substance in our body and the foods we eat further validates this perspective. Neither the National Collegiate Athletic Association nor the International Olympic Committee—both very strict organizations when it comes to regulating supplement use—prohibits Cr supplementation.

The principal benefits of Cr supplementation to athletes include increases in strength, speed, and muscle mass—effects that occur secondary to the athlete's ability to train longer and at a higher intensity and recover quicker because of enhanced cellular energy provisions. More recently, Cr has shown promise in improving general health. There is research to support a role for Cr supplementation in the treatment of muscular dystrophy and Lou Gehrig's disease. Cr supplementation decreases the level of homocysteine—a contributing factor to cardiovascular disease—in the blood, and the supplement might also possess anti-inflammatory and antioxidant properties. Cr is best taken postexercise with carbohydrates, and contrary to popular belief, it is unnecessary to "load" Cr. A smaller daily dose (3–5 g) over several weeks is equally effective in increasing muscle-Cr stores with a decreased likelihood of gastrointestinal problems (the most common side effect) as stated by Eric Cressey, M.A., C.S.C.S.

4.2. Side Bar No. 2

4.2.1. Carbohydrate and Cr Loading

The concept of nonresponders and responders to Cr supplementation might be eliminated if large doses of carbohydrate are ingested with large doses of Cr for an initial 5-d "loading phase." There have been a few investigations that have shown how ingestion of a large amount of glucose along with Cr supplementation during the loading phase reduced variability in muscle-Cr uptake and enhanced Cr accumulation. The enhancement of Cr accumulation in muscle when combined with the simple carbohydrate glucose is thought to be a result of carbohydrate-mediated insulin release. Insulin has been shown to stimulate muscle-Cr transport through the membrane transport protein, CreaT. However, the investigations that suggest large doses of carbohydrate are required to enhance Cr uptake have used very-high doses of carbohydrate, such as 370 g/d, which adds an extra 1500 kcals to the total-energy intake. For athletes and recreational exercisers looking to enhance their performance with Cr, taking such a high dose of pure glucose may not be advisable or even palatable. It also might compromise health and body composition. Other researchers have shown that lower doses of carbohydrate taken with Cr, such as 1 g of both carbohydrate and Cr per kg bodyweight taken twice per day can also improve Cr uptake, relative to taking Cr alone. However, this is still a large dose of carbohydrate (~100 g) and may not be desirable or economical for any person to ingest. As an alternative, people may still enhance their Cr stores by eliminating the large carbohydrate and Cr dose required in the loading phase, but waiting longer for the ergogenic effects to be noticed. Therefore, if one wishes to enhance their muscle-Cr stores without consuming a large dose of carbohydrate, they could supplement slowly, avoiding the loading phase, and consume a small dose of carbohydrate to slightly increase insulin and enhance Cr uptake.

REFERENCES

1. Chanutin A, Guy LP. The fate of creatine when administered to man. J Biol Chem 1926; 67:29–41.
2. Wyss M, Kaddurah-Daouk R. Creatine and creatinine metabolism. Physiol Rev 2000; 80:1107–1213.
3. Walker JB. Creatine: biosynthesis, regulation, and function. Adv Enzymol Relat Areas Mol Biol 1979; 50:177–242.
4. Hoberman HD, Sims EAH, Engstrom WW. The effect of methyltestosterone on the rate of synthesis of creatine. J Biol Chem 1948; 173:111–116.
5. Verhoeven NM, Salomons GS, Jakobs C. Laboratory diagnosis of defects of creatine biosynthesis and transport. Clin Chim Acta 2005; 361:1–9.
6. Fatterpaker P, Marfatia U, Sreenivasan A. Influence of folic acid and vitamin B12 on formation of creatine in vitro and in vivo. Nature 1951; 167:1067–1068.
7. Stekol JA, Weiss S, Smith P, Weiss K. The synthesis of choline and creatine in rats under various dietary conditions. J Biol Chem 1953; 201:299–316.
8. Hoogwerf BJ, Laine DC, Greene E. Urine C-peptide and creatinine (Jaffe method) excretion in healthy young adults on varied diets: sustained effects of varied carbohydrate, protein, and meat content. Am J Clin Nutr 1986; 43:350–360.
9. Harris RC, Nevill M, Harris DB, Fallowfield JL, Bogdanis GC, Wise JA. Absorption of creatine supplied as a drink, in meat or in solid form. J Sports Sci 2002; 20:147–151.
10. Vandenberghe K, Goris M, Van Hecke P, Van Leemputte M, Vangerven L, Hespel P. Long-term creatine intake is beneficial to muscle performance during resistance training. J Appl Physiol 1997; 83:2055–2063.
11. Rossiter HB, Cannell ER, Jakeman PM. The effect of oral creatine supplementation on the 1000-m performance of competitive rowers. J Sports Sci 1996; 14:175–179.
12. Harris RC, Soderlund K, Hultman E. Elevation of creatine in resting and exercised muscle of normal subjects by creatine supplementation. Clin Sci (Lond) 1992; 83:367–374.

13. Guerrero-Ontiveros ML, Wallimann T. Creatine supplementation in health and disease. Effects of chronic creatine ingestion in vivo: down-regulation of the expression of creatine transporter isoforms in skeletal muscle. Mol Cell Biochem 1998; 184:427–437.
14. Loike JD, Zalutsky DL, Kaback E, Miranda AF, Silverstein SC. Extracellular creatine regulates creatine transport in rat and human muscle cells. Proc Natl Acad Sci USA 1988; 85:807–811.
15. Snow RJ, Murphy RM. Creatine and the creatine transporter: a review. Mol Cell Biochem 2001; 224:169–181.
16. Balsom PD, Soderlund K, Sjodin B, Ekblom B. Skeletal muscle metabolism during short duration high-intensity exercise: influence of creatine supplementation. Acta Physiol Scand 1995; 154:303–310.
17. Casey A, Constantin-Teodosiu D, Howell S, Hultman E, Greenhaff PL. Creatine ingestion favorably affects performance and muscle metabolism during maximal exercise in humans. Am J Physiol 1996; 271:E31–E37.
18. Meyer RA, Brown TR, Kushmerick MJ. Phosphorus nuclear magnetic resonance of fast- and slow-twitch muscle. Am J Physiol 1985; 248:C279–C287.
19. Kushmerick MJ, Moerland TS, Wiseman RW. Mammalian skeletal muscle fibers distinguished by contents of phosphocreatine, ATP, and Pi. Proc Natl Acad Sci USA 1992; 89:7521–7525.
20. Green AL, Hultman E, Macdonald IA, Sewell DA, Greenhaff PL. Carbohydrate ingestion augments skeletal muscle creatine accumulation during creatine supplementation in humans. Am J Physiol 1996; 271:E821–E826.
21. Hultman E, Soderlund K, Timmons JA, Cederblad G, Greenhaff PL. Muscle creatine loading in men. J Appl Physiol 1996; 81:232–237.
22. Maughan RJ. Creatine supplementation and exercise performance. Int J Sport Nutr 1995; 5:94–101.
23. Lukaszuk JM, Robertson RJ, Arch JE, et al. Effect of creatine supplementation and a lacto-ovo-vegetarian diet on muscle creatine concentration. Int J Sport Nutr Exerc Metab 2002; 12:336–348.
24. Snow RJ, Murphy RM. Factors influencing creatine loading into human skeletal muscle. Exerc Sport Sci Rev 2003; 31:154–158.
25. Virgili F, Maiani G, Zahoor ZH, Ciarapica D, Raguzzini A, Ferro-Luzzi A. Relationship between fat-free mass and urinary excretion of creatinine and 3-methylhistidine in adult humans. J Appl Physiol 1994; 76:1946–1950.
26. Bogdanis GC, Nevill ME, Boobis LH, Lakomy HK. Contribution of phosphocreatine and aerobic metabolism to energy supply during repeated sprint exercise. J Appl Physiol 1996; 80:876–884.
27. Greenhaff PL, Bodin K, Soderlund K, Hultman E. Effect of oral creatine supplementation on skeletal muscle phosphocreatine resynthesis. Am J Physiol 1994; 266:E725–E730.
28. van Leemputte M, Vandenberghe K, Hespel P. Shortening of muscle relaxation time after creatine loading. J Appl Physiol 1999; 86:840–844.
29. McArdle WD, Frank I. Katch, Victor L. Katch. Sports and Exercise Nutrition. 2nd ed. Lippincott, Williams, and Wilkins, Baltimore, 2005.
30. Bessman SP, Geiger PJ. Transport of energy in muscle: the phosphorylcreatine shuttle. Science 1981; 211:448–452.
31. Ogut O, Brozovich FV. Creatine phosphate consumption and the actomyosin crossbridge cycle in cardiac muscles. Circ Res 2003; 93:54–60.

32. Meyer RA, Sweeney HL, Kushmerick MJ. A simple analysis of the "phosphocreatine shuttle". Am J Physiol 1984; 246:C365–C377.
33. Thompson CH, Kemp GJ, Sanderson AL, et al. Effect of creatine on aerobic and anaerobic metabolism in skeletal muscle in swimmers. Br J Sports Med 1996; 30:222–225.
34. Volek JS, Duncan ND, Mazzetti SA, et al. Performance and muscle fiber adaptations to creatine supplementation and heavy resistance training. Med Sci Sports Exerc 1999; 31:1147–1156.
35. Green AL, Simpson EJ, Littlewood JJ, Macdonald IA, Greenhaff PL. Carbohydrate ingestion augments creatine retention during creatine feeding in humans. Acta Physiol Scand 1996; 158:195–202.
36. Steenge GR, Simpson EJ, Greenhaff PL. Protein- and carbohydrate-induced augmentation of whole body creatine retention in humans. J Appl Physiol 2000; 89:1165–1171.
37. Preen D, Dawson B, Goodman C, Beilby J, Ching S. Creatine supplementation: a comparison of loading and maintenance protocols on creatine uptake by human skeletal muscle. Int J Sport Nutr Exerc Metab 2003; 13:97–111.
38. Robinson TM, Sewell DA, Hultman E, Greenhaff PL. Role of submaximal exercise in promoting creatine and glycogen accumulation in human skeletal muscle. J Appl Physiol 1999; 87:598–604.
39. Zange J, Kornblum C, Muller K, et al. Creatine supplementation results in elevated phosphocreatine/adenosine triphosphate (ATP) ratios in the calf muscle of athletes but not in patients with myopathies. Ann Neurol 2002; 52:126, author reply 126, 127.
40. Murphy RM, Tunstall RJ, Mehan KA, et al. Human skeletal muscle creatine transporter mRNA and protein expression in healthy, young males and females. Mol Cell Biochem 2003; 244:151–157.
41. Parise G, Mihic S, MacLennan D, Yarasheski KE, Tarnopolsky MA. Effects of acute creatine monohydrate supplementation on leucine kinetics and mixed-muscle protein synthesis. J Appl Physiol 2001; 91:1041–1047.
42. Persky AM, Brazeau GA. Clinical pharmacology of the dietary supplement creatine monohydrate. Pharmacol Rev 2001; 53:161–176.
43. Rawson ES, Gunn B, Clarkson PM. The effects of creatine supplementation on exercise-induced muscle damage. J Strength Cond Res 2001; 15:178–184.
44. Francaux M, Poortmans JR. Effects of training and creatine supplement on muscle strength and body mass. Eur J Appl Physiol Occup Physiol 1999; 80:165–168.
45. Low SY, Rennie MJ, Taylor PM. Modulation of glycogen synthesis in rat skeletal muscle by changes in cell volume. J Physiol 1996; 495(Pt 2):299–303.
46. Derave W, Eijnde BO, Verbessem P, et al. Combined creatine and protein supplementation in conjunction with resistance training promotes muscle GLUT-4 content and glucose tolerance in humans. J Appl Physiol 2003; 94:1910–1916.
47. Nelson AG, Arnall DA, Kokkonen J, Day R, Evans J. Muscle glycogen supercompensation is enhanced by prior creatine supplementation. Med Sci Sports Exerc 2001; 33:1096–1100.
48. Op 't Eijnde B, Urso B, Richter EA, Greenhaff PL, Hespel P. Effect of oral creatine supplementation on human muscle GLUT4 protein content after immobilization. Diabetes 2001; 50:18–23.
49. van Loon LJ, Murphy R, Oosterlaar AM, et al. Creatine supplementation increases glycogen storage but not GLUT-4 expression in human skeletal muscle. Clin Sci (Lond) 2004; 106:99–106.

50. Maughan RJ, Poole DC. The effects of a glycogen-loading regimen on the capacity to perform anaerobic exercise. Eur J Appl Physiol Occup Physiol 1981; 46:211–219.
51. Pizza FX, Flynn MG, Duscha BD, Holden J, Kubitz ER. A carbohydrate loading regimen improves high intensity, short duration exercise performance. Int J Sport Nutr 1995; 5:110–116.
52. Haff G, Stone MH, Warren BJ, et al The effect of carbohydrate supplementation on multiple sessions and bouts of resistance exercise. J Strength Cond Res 1999; 13:111.
53. Haff GG, Koch AJ, Potteiger JA, et al. Carbohydrate supplementation attenuates muscle glycogen loss during acute bouts of resistance exercise. Int J Sport Nutr Exerc Metab 2000; 10:326–339.
54. Kreider RB, Ferreira M, Wilson M, et al. Effects of creatine supplementation on body composition, strength, and sprint performance. Med Sci Sports Exerc 1998; 30:73–82.
55. Hespel P, Op't Eijnde B, Van Leemputte M, et al. Oral creatine supplementation facilitates the rehabilitation of disuse atrophy and alters the expression of muscle myogenic factors in humans. J Physiol 2001; 536:625–633.
56. Willoughby DS, Rosene J. Effects of oral creatine and resistance training on myosin heavy chain expression. Med Sci Sports Exerc 2001; 33:1674–1681.
57. Willoughby DS, Rosene JM. Effects of oral creatine and resistance training on myogenic regulatory factor expression. Med Sci Sports Exerc 2003; 35:923–929.
58. Buckingham M, Bajard L, Chang T, et al. The formation of skeletal muscle: from somite to limb. J Anat 2003; 202:59–68.
59. Rosenblatt JD, Parry DJ. Adaptation of rat extensor digitorum longus muscle to gamma irradiation and overload. Pflugers Arch 1993; 423:255–264.
60. Barton-Davis ER, Shoturma DI, Sweeney HL. Contribution of satellite cells to IGF-I induced hypertrophy of skeletal muscle. Acta Physiol Scand 1999; 167:301–305.
61. Dangott B, Schultz E, Mozdziak PE. Dietary creatine monohydrate supplementation increases satellite cell mitotic activity during compensatory hypertrophy. Int J Sports Med 2000; 21:13–16.
62. Vierck JL, Icenoggle DL, Bucci L, Dodson MV. The effects of ergogenic compounds on myogenic satellite cells. Med Sci Sports Exerc 2003; 35:769–776.
63. Olsen S, Aagaard P, Kadi F, et al. Creatine supplementation augments the increase in satellite cell and myonuclei number in human skeletal muscle induced by strength training. J Physiol 2006; 573:525–534.

2 Creatine Supplementation in Strength-Power Sports

Darryn S. Willoughby, PhD

1. INTRODUCTION

The exogenous ingestion of creatine (Cr) is typically used as a performance enhancing (ergogenic) supplement because it is known to improve performance in muscular strength and power activities, enhance short bursts of muscular endurance, and allow for greater muscular overload in order to improve training effectiveness. Creatine has become one of the most popular ingested nutritional supplements due to its potential enhancement of athletic performance. Creatine is primarily located in skeletal muscle and plays a pivotal role in cellular bioenergetics, specifically towards the reformation of a molecule essential for muscular contraction, adenosine triphosphate (ATP). The vast majority of research indicates that high-intensity, short duration, and repeated exercise bouts are the most effective modes of exercise that can be enhanced by creatine supplementation. Oral creatine supplementation has been shown to provide numerous benefits, including increases in lean muscle mass, muscular strength, and enhanced performance in various athletic capacities. The creatine transporter is a protein that mediates the entry of creatine from the circulation into the muscle cell.

2. CREATINE'S ROLE IN MUSCULAR STRENGTH AND POWER

Numerous research studies have demonstrated that supplemental creatine (Cr) monohydrate is an effective nutritional compound in increasing short-term (1–3 wk) and long-term (4–12 wk) muscle performance, as indicated by an increase in strength and power. In addition,

From: *Essentials of Creatine in Sports and Health*
Edited by: J. R. Stout, J. Antonio and D. Kalman © Humana Press Inc., Totowa, NJ

supplemental Cr appears to also be a beneficial catalyst in increasing muscle hypertrophy and total body mass when combined with resistance training. Based on the overwhelming number of studies conducted in the last 10 yr that have demonstrated that Cr supplementation improves exercise performance and/or training adaptations, Cr has quickly become one of the most popular ergogenic sport supplements used today. Most of the studies with Cr supplementation have used male participants. However, about one-third of the studies have evaluated women and/or mixed cohorts of men and women and have demonstrated women to undergo ergogenic benefits following Cr supplementation. Although, gains in body mass, fat-free mass, and muscle strength and power are generally not as rapid as men, these studies do suggest that women do benefit from Cr supplementation.

However, one must consider that the impact of Cr's effectiveness in increasing muscle strength and power is predicated on the duration and dosage of Cr supplementation, as well as the muscle-Cr transport and uptake capacity. Cr is typically ingested at an approximate dosage of 20 g/d (0.3 g/kg body mass/d) for 5–7 d during a loading phase and then around 5 g/d for several weeks during a maintenance phase. The majority of Cr uptake during a loading phase typically occurs during the first 2–3 d of the loading period, and research has shown that the most rapid way to increase muscle-Cr stores is to use a loading method. However, there are studies that have reported that 5–6 g/d of Cr supplementation for 10–12 wk promoted greater gains in strength and muscle mass during training when compared with placebo. In addition, a 4-wk study that did not use a loading phase, but rather provided a maintenance dose of 3 g/d for the duration of the study, still showed significant Cr-induced performance benefits. However, in light of this, using a loading phase and then using a maintenance dose of 5–6 g/d (0.07 g/kg body mass/d) appears to be necessary to maintain Cr stores in most individuals owing to the fact that larger individuals are likely to retain 2–4 g/d of Cr after loading periods, if adequate Cr is ingested during the maintenance phase.

Evidence suggests that Cr supplementation is more likely suited to improve performance in the sports/activities that require more of an anaerobic performance component *(1)* such as weight lifting, sprinting, football, and ice hockey. Cr enhances short-term, anaerobic endurance through its inherent ability to enhance muscle bioenergetics. In addition, Cr indirectly promotes muscle anabolism (growth) through long-term supplementation coupled with resistance training by extending exercise output, again through enhanced muscle bioenergetics. As a result, muscles then compensate for the increased mechanical load through the production

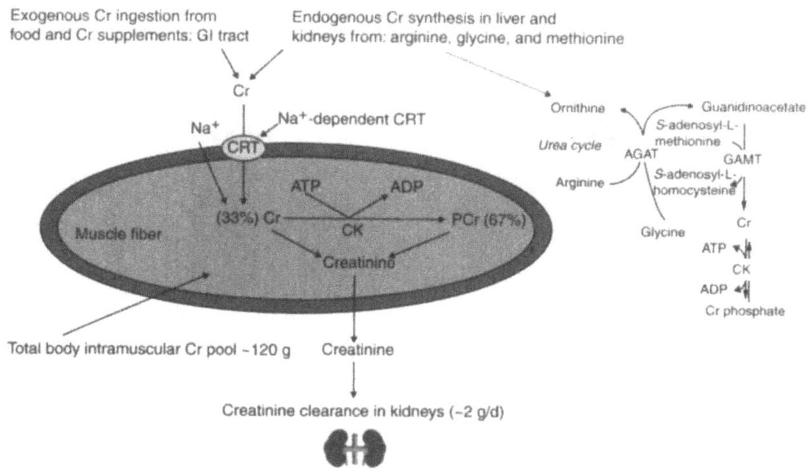

Fig. 1. Illustration of the process of intramuscular-Cr uptake and creatinine clearance, in addition to the bioenergetic mechanism in which Cr provides energy substrate.

of new muscle proteins. These newly added proteins promote hypertrophy, thereby allowing muscles to generate greater amounts of force and power.

However, for sports requiring mainly aerobic performance there is less evidence that Cr supplementation is helpful. So, for endurance athletes such as runners, cyclists, and long-distance swimmers, many unknowns still exist. Nevertheless, a few studies have shown some improvement in performance. For example, a study found that Cr supplements delayed the onset of muscle fatigue in endurance athletes by boosting their lactate thresholds *(2)*.

3. BIOENERGETIC MECHANISMS FOR STRENGTH/POWER INCREASES WITH CR SUPPLEMENTATION

Cr, or methyl guanidine-acetic acid, is a naturally occurring compound endogenously synthesized from arginine, glycine, and methionine. To meet the demands of a high-intensity exercise, such as a sprint, muscles derive their energy from a series of reactions involving ATP, phosphocreatine (PCr), ADP, and Cr. Once in the bloodstream, Cr is then taken up by muscle fibers predominately by way of a sodium chloride (Na^+/Cl^-)-dependent Cr transporter (CRT). The major rationale of Cr supplementation is to maximize the intracellular pool of total-Cr (Cr + PCr). Ingestion of Cr monohydrate at a rate of 20 g/d for 5–6 d increases total-Cr concentration by approx 30% as PCr *(3)* (Fig. 1).

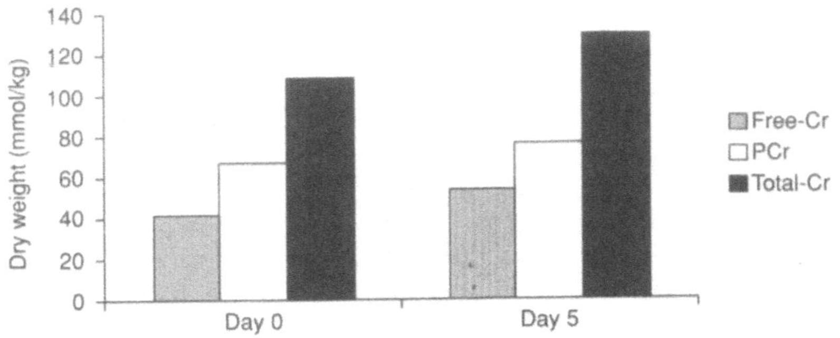

Fig. 2. Five days of Cr loading at 20 g/d increases intramuscular levels of free-Cr, PCr, and total-Cr by 30, 17, and 20%, respectively. Adapted from ref. *4*.

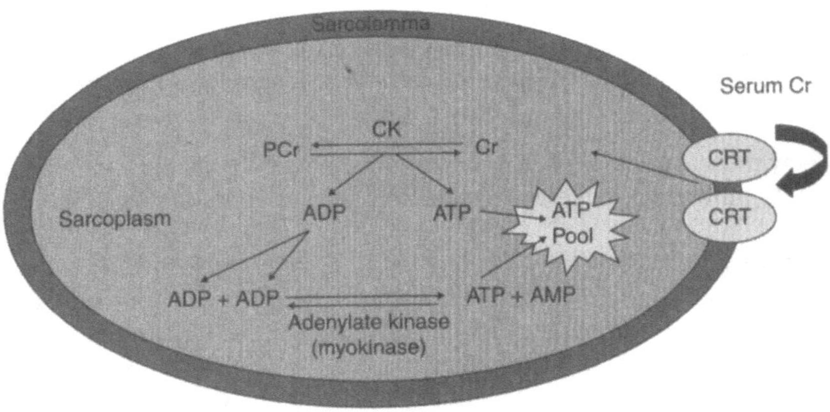

Fig. 3. Illustration of the intramuscular mechanisms of ATP synthesis by way of PCr as substrate and the ability of the ADP produced from Cr phosphorylation to be used to resynthesize ATP by way of the enzymatic activation of myokinase.

Figure 2 illustrates data demonstrating that 5 d of Cr supplementation at a dose of 20 g/d increased the content of free-Cr, PCr, and total-Cr by 30, 17, and 20%, respectively *(4)*.

Through the depletion of intracellular-PCr stores, the intracellular concentration of ATP, a vital molecule necessary for muscle contraction, is maintained and replenished. This occurs through a freely reversible reaction in which PCr phosphorylates ADP to replenish ATP stores, catalyzed through the enzyme Cr kinase (CK). As illustrated in Fig. 3, ATP is regenerated when PCr donates a phosphate molecule that combines with ADP. Stored PCr can fuel the first 4–5 s of a sprint, but another fuel source must provide the energy to sustain the activity, such as the ATP synthesized by way of the enzymatic activation of adenylate

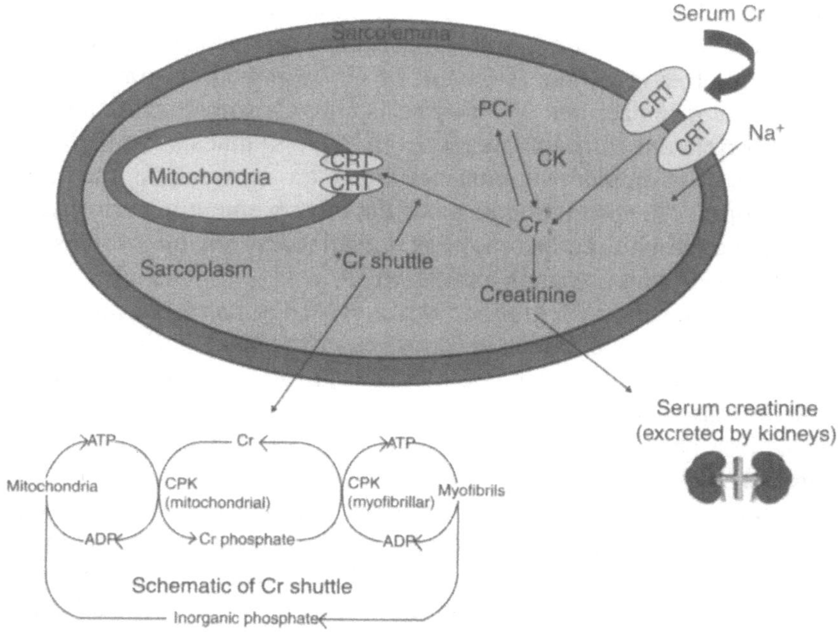

Fig. 4. Illustration of the process of intramuscular-Cr uptake involving the sarcolemmal and mitochondrial CRT, and the associated role of the Cr shuttle in the oxidative synthesis of ATP.

kinase (myokinase) utilizing ADP as substrate (*see* Fig. 4). However, Cr supplements increase the storage of PCr, thus making more ATP available to fuel the working muscles and enable them to work harder before becoming fatigued.

PCr levels within the muscle are almost three to four times more abundant than intramuscular-ATP stores. Although, PCr is more copious than ATP, the rate in which ATP is utilized is likely to exceed the overall energy substrate regeneration necessary at activities of high intensity. However, the PCr supply is sufficient in providing a temporary ATP source until other bioenergetic systems reach maximal rates. Furthermore, there is much evidence indicating that Cr supplementation can improve athletic performance and myocellular bioenergetics. This is important because the intracellular concentration of PCr plays a significant role during the immediate bioenergetic system, which is most active during exercise at high intensity, short duration, and repeated bouts of physical activity.

Increasing intramuscular levels of Cr and PCr can affect muscle strength and power by increasing the availability of PCr, thereby enhancing the availability of energy substrate during high-intensity exercise like sprinting and intense weightlifting. For instance, forearm muscles of males were examined during a 10-s maximal dynamic handgrip exercise before and after ingestion of 30 g/d of Cr monohydrate for 14 d. An increase in total anaerobic ATP synthesis after Cr supplementation positively correlated with the increase in ATP synthesis through PCr hydrolysis. Cr supplementation produced an increase in the mean power output. The results suggest that the improvement in performance was associated with the increased PCr availability for the synthesis of ATP resulting from Cr supplementation *(5)*.

Cr supplemented at 25 g/d for 5 d was shown to increase muscle-PCr levels by 16% and was shown to improve performance during rapid and dynamic intermittent muscle contractions; however, Cr loading did not facilitate muscle-PCr resynthesis during intermittent isometric muscle contractions *(6)*. Furthermore, elevated levels of PCr are likely to help improve recovery time between sprints and/or bouts of intense exercise. In summary, although all energy systems, both aerobic and anaerobic, will at least partially utilize the phosphagen system, high-intensity, short-duration, and repeated exercise bouts have been repeatedly observed in the literature to be the most effective mode of exercise that can be enhanced through Cr supplementation *(7)*.

4. ROLE OF THE CRT IN REGULATING MUSCLE-CR UPTAKE

As a result of the various mechanisms in which Cr appears to be involved, the regulation of total-Cr metabolism within the muscle is still poorly understood. Whole-body Cr retention is dependent primarily on rates of Cr uptake and intramuscular-Cr content, and to a lesser extent, the slow degradation of Cr into creatinine. Cr uptake into the muscle is dependent on the CRT, a membrane-spanning protein that transfers Cr from the blood into the muscle fibers. It is likely that regulation of the CRT protein is important in controlling intramuscular-Cr levels *(8)*. This becomes apparent in certain Cr-deficient pathologies in which the CRT might be defective or absent, such that supplementation is unable to restore Cr to ordinary levels *(9)*. Strangely, chronic Cr supplementation has been demonstrated to cause a reduction of the CRT protein in rats, whereas chronic supplementation of β-guanidinoproprionic acid, a Cr analog that competitively inhibits the CRT, resulted in an increase in CRT protein *(10)*. This suggests that intramuscular-Cr content might regulate the amount of CRT protein present in muscle. More recently, Cr transport has been identified in the mitochondria *(11)*, which suggests that Cr may not only exist in the sarcoplasm of the muscle fibers, but that there might be an intermitochondrial pool of Cr as well.

All plasma membranes are impermeable to the diffusion of Cr and PCr. More than 90% of intramuscular Cr is shuttled across the sarcolemma against a concentration gradient by the CRT. Approximately 65% of the Cr that enters skeletal muscle will be actively transphosphorylated by CK to produce PCr, such that it is readily available to react to ATP depletion *(12)*. PCr cannot be transported by CRT, and as such, CK renders a portion of the intracellular-Cr pool to be confined in the cytosol.

Recently, a new and possibly unique CRT activity in the mitochondria has been discovered, which suggests that there are likely two separate pools of intracellular Cr, regulated by different CRT activities *(13)*. It is possible that these separate CRTs are entirely different proteins, with variations in structure and subject to different degrees of regulation. It is currently uncertain what purpose a mitochondrial pool of Cr may have, although it might simply serve as a buffer, should plasma-Cr levels drop appreciably, such that cytosolic-Cr levels might remain elevated during high-contractile activity. Acute regulation (within hours) of CRT activity might be directly influenced by fluctuations in plasma-Cr concentrations. This may involve alterations in the flux of sodium across the sarcolemma *(14)*.

Similarly, there may be factors that might directly stimulate or inhibit the CRT protein, and chronic adaptations (days to weeks) might involve altering the number of CRT proteins available at the membrane, or by altering the number of CRT proteins expressed by the cell *(15)*. It seems that Cr uptake is actually inhibited with prolonged exposure to high plasma-Cr levels, which might be because of decreased activity of CRT. This suggests that Cr uptake is actually dependent on intracellular-Cr concentrations, and not extracellular-Cr concentrations. It appears that elevated plasma-Cr levels promote an initial rise in Cr uptake and resultant intracellular-Cr concentration, which may by itself begin to inhibit uptake by negative feedback. This downregulation of Cr uptake with chronic elevated plasma-Cr levels may be a result of an inhibitory protein and suggests that high-intracellular concentrations may induce the expression of a protein that might either directly inhibit the CRT protein, or somehow reduce the number of transporter proteins available at the membrane.

5. TRANSPORT OF ENERGY IN MUSCLE: THE CR SHUTTLE

Exercise results in the production of Cr in contracting muscle fibers. This might not be entirely dependent on the sacroplasmic activity of CK, and serves to impart oxidative/respiratory control on the muscle mitochondria. Because of this functional compartmentalization of CK on the mitochondrion, it is clear that the actual form of energy transport in the muscle fiber is PCr. The finding of an isoenzyme of CK attached to the M-line region of the myofibril revealed that the peripheral receptor for PCr generated within the mitochondria is the molecular basis for a Cr phosphate shuttle for energy transport in skeletal muscle, and thereby demonstrates a direct relation between muscle activity and the concentrations of intramuscular ATP and ADP levels *(10)*. The adenine nucleotide charge regulates both activity of glycolysis and the tricarboxylic acid cycle, and is inherently linked to the sacroplasmic activity of CK, and the subsequent levels of PCr vs Cr (*see* Fig. 1). As a result, the ratio between ADP and ATP has the ability to stimulate and inhibit oxidative phosphorylation. During exercise, an increase in ADP occurs relative to ATP and is associated with an increase in whole-body oxygen consumption. However, when exercise stops, the intracellular oxidative mechanisms soon re-establish normal levels of ATP, ADP, and adenosine monophosphate.

Consequently, whole-body oxygen consumption rapidly declines toward resting levels after exercise. The Cr phosphate shuttle is the mechanism in which ATPase activity of the contractile machinery is buffered by sarcoplasmic PCr, with the ultimate result being phosphorylation of mitochondrial ADP to ATP and appears to be a significant oxidative mechanism which supports sustained, endurance exercise. Therefore, the Cr phosphate shuttle may not be a significant factor involved in the bioenergetic processes in which short-term Cr supplementation can effectively increase muscle strength and power during high-intensity muscular efforts. As a result, it certainly becomes more an issue with the increase in muscle strength, hypertrophy, or endurance capacity during long-term Cr supplementation associated with resistance-training or other forms of sport-specific training.

6. HYPERTROPHIC MECHANISMS FOR STRENGTH/POWER INCREASES WITH CR SUPPLEMENTATION

Early Cr supplementation research primarily documented increases in short-term muscle strength and power, and these improvements were likened to Cr's ability to enhance intramuscular-phosphagen levels and bioenergetic coupling of ATP resynthesis. When incorporated into a long-term resistance training program this might allow an athlete to do more work over a series of sprints and/or sets of exercise, and over time may subsequently lead to greater gains in strength, muscle mass, and performance. Although, enhanced myocellular bioenergetics associated with Cr supplementation appear to be the primary mechanisms for increasing short-term strength and power, they cannot completely account for increases in muscle strength and power over the course of a long-term resistance training program that appear to also occur by way of hypertrophic mechanisms.

By and large, there has been considerably more research done regarding the short-term effects of Cr supplementation on muscle strength and power. However, this is typically not how many athletes utilize Cr in their training regimens and it became quite evident that data was needed highlighting any performance benefits with long-term Cr supplementation. Therefore, as studies began to occur, long-term resistance training programs combined with Cr supplementation began to document increases in muscle strength and total body mass and speculated that Cr supplementation had a positive effect on muscle hypertrophy.

In one of the first long-term studies, no loading phase was used but rather 5 g of Cr was ingested daily for 10 wk, along with resistance training 4 d/wk. Cr supplementation was shown to preferentially increase indices of muscle strength and power, whereas increasing total body mass without any concomitant increases in fat mass. As promising as this study was in suggesting a possible increase in muscle hypertrophy, no direct measurements of muscle hypertrophy was performed *(16)*.

In a classic study, Cr's ability to increase muscle strength and total body mass primarily as the result of a hypertrophic mechanism was illustrated. During this study, a 1-wk Cr-loading phase at 20 g/d was followed by 11 wk of 5 g/d in conjunction with resistance training performed on an average of thrice weekly. Results showed that Cr preferentially increased muscle strength, fat-free mass, and types I, IIa, and IIab muscle fibers *(17)*.

However, the first study to actually demonstrate a direct hypertrophic link to Cr supplementation through molecular mechanisms linked to both muscle-specific gene and protein expression was published in 2001 *(18)*. This study involved 12 wk of resistance training, thrice weekly, combined with the daily ingestion of 6 g/d of Cr monohydrate (without a loading phase), and found that Cr preferentially increased muscle strength, fat-free mass, thigh mass, and types I and IIa myosin heavy chain isoform mRNA and protein expression, and myofibrillar protein content, and suggests that the enhancements in muscle mass and performance from Cr occur by way of a molecular, hypertrophic mechanism.

In an attempt to determine a possible pre- or post-translational hypertrophic mechanism to help explain Cr's presumable hypertrophic role, a follow-up study demonstrated that resistance training increased the mRNA and protein expression of the myogenic regulatory factor, Myo-D, whereas Cr preferentially increased the mRNA and protein expression of the myogenic regulatory factors myogenin and myogenic

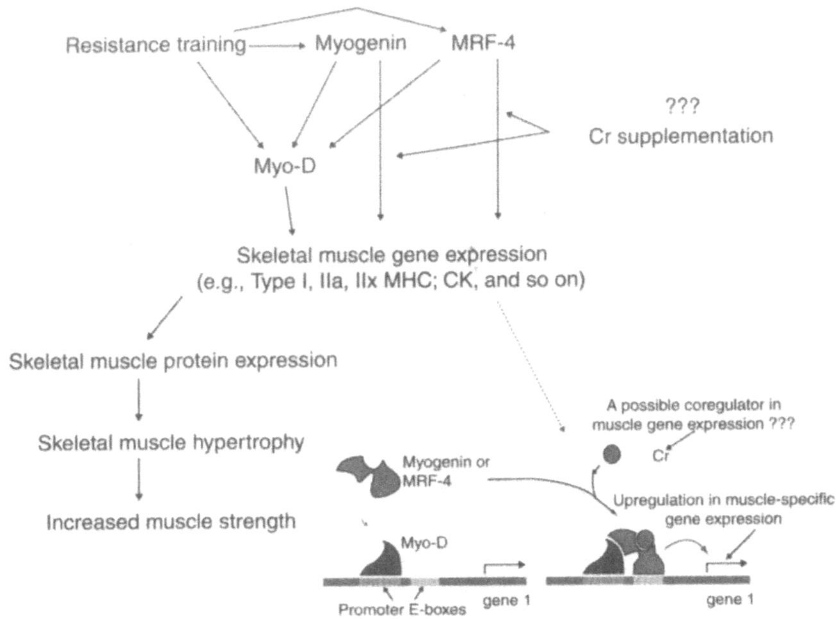

Fig. 5. Based on previous studies, an illustration of the theoretical mechanism in which resistance training combined with Cr supplementation might interact with the myogenic regulatory factors MRF-4 and myogenin, thereby resulting in a preferential increase in muscle-specific gene expression and subsequently increasing muscle protein expression, hypertrophy, and muscle strength *(18–20)*.

regulatory factor (MRF)-4, DNA-binding proteins that serve as transcription factors in upregulating the expression of muscle-specific genes. Additionally, myogenin and MRF-4 expression was directly correlated to the mRNA expression of the skeletal muscle isozyme of CK. This study indicates Cr to operate as a possible transcriptional cofactor, working in concert with the myogenic regulatory factors Myo-D, MRF-4, and myogenin in regulating the expression of selected muscle-specific genes *(19)* (Fig. 5).

Also, research has shown that Cr increases the activity of myogenic cells. These cells, sometimes called satellite cells, are myogenic stem cells that are involved in hypertrophy of adult skeletal muscle. Satellite cells reside in a quiescent state within the sarcolemma until various stressors such as resistance exercise and/or muscle injury upregulates their activity. Following proliferation and subsequent differentiation, these satellite cells will fuse with one another or with the adjacent damaged muscle fiber, thereby increasing myonuclei numbers necessary for fiber growth and repair.

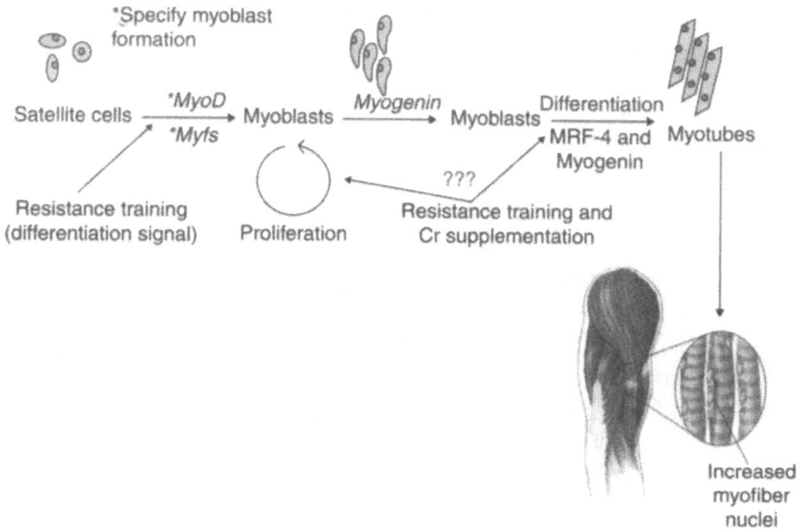

Fig. 6. Based on previous studies, an illustration of the theoretical mechanism in which resistance training combined with Cr supplementation might interact with the myogenic regulatory factors MRF-4 and myogenin, thereby resulting in satellite cell proliferation and differentiation and subsequently increasing satellite cell number and myonuclei content *(20,21)*.

A study was conducted showing that after 2 wk of immobilization followed by 10 wk of Cr supplementation (20 g/d for 1 wk followed by 5 g/d for 9 wk) and resistance training thrice weekly increased muscle fiber diameter and MRF-4 protein expression, and that the increase in MRF-4 expression was correlated with the increase in muscle fiber diameter. These data suggest that Cr increased the number of myonuclei donated from satellite cells, and that this increase in myonuclei probably stems from Cr's ability to increase levels of the myogenic transcription factor MRF-4 *(20)*. A more recent study has shown that three weekly resistance training sessions for 15 wk combined with the daily ingestion of 6 g/d of Cr showed increases in muscle strength, muscle fiber size, satellite cell number, and muscle myonuclei content. Additionally, the increase in muscle-fiber size was associated with the myonuclei content *(21)*. This study is the first to indicate that long-term Cr supplementation combined with resistance training produces increases in muscle strength hypertrophy, primarily because of an increase in satellite cell activation. However, the mechanism in which Cr apparently preferentially activated satellite cells is presently unknown (Fig. 6).

7. EFFECT OF CR SUPPLEMENTATION ON MUSCLE STRENGTH AND POWER

Of the hundreds of studies examining Cr supplementation, more than 80% prove statistically significant toward short- and/or long-term performance enhancement. Most of these studies have been performed in untrained participants; however, some have been conducted in competitive athletes. Even then, several well-controlled studies have demonstrated Cr supplementation not to have a positive effect on muscular performance. Although many Cr studies differ in regard to the duration and dosage of Cr supplementation, the specific Cr formulation, the age, gender, and training status of the participants, methods of performance assessment, and other differences in experimental design, one should be also aware of the fact that all individuals are not likely to have similar responses to Cr supplementation.

7.1. Short-Term Performance Effects

Short-term Cr supplementation appears to work best during high-intensity exercise like repetitive sprinting and weightlifting. Most studies indicate that Cr supplementation increases body mass by about 1–2 kg in the first week of loading. However, Cr is an osmotically active compound, such that increases in total body Cr retention caused by oral Cr supplementation should result in concomitant increases in water retention. This in part, might explain how increases in total body mass are seen after only a loading supplementation of 20–25 g/d for 5–7 d *(22)*. In particular, it would be expected that increases in intramuscular Cr, which accounts for approx 95% of total body Cr, would particularly augment intramuscular water volume. However, it has been shown that a loading phase of 25 g/d for 7 d followed by a maintenance phase of 5 g/d for 21 d resulted in increases in intramuscular Cr and total body water, without altering proportional fluid distribution *(23)*. Interestingly, increases in cell volume appear to be a proliferative, anabolic signal that may enhance protein synthesis *(24)*, which suggests a method by which extended Cr supplementation might promote muscular hypertrophy with long-term Cr supplementation.

In individuals confined to laboratory settings involving exercises such as bicycling or resistance exercise rather than sport training, Cr's role in increasing short-term muscle performance has been documented. The effect of short-term Cr supplementation on high-intensity exercise performance in males and females was determined by providing 5 g/d of Cr monohydrate for 4 d. Cr significantly increased peak and relative peak anaerobic cycling power and maximal strength with no gender-specific responses, suggesting that short-term Cr supplementation can increase indices of high-intensity exercise performance for both males and females *(25)*. In addition, the effect of Cr supplementation on muscle power during repeated bouts of supramaximal bicycling exercise was performed five times with 2-min intervals. Male subjects ingested 5 g/d of Cr monohydrate for 6 d. Cr increased both mean and total power output; this suggests that Cr supplementation enhances total power output during the repeated bouts of supramaximal exercise separated by short resting intervals *(26)*. The effects of Cr supplementation on force generation during an isometric bench-press in resistance-trained men was determined by having subjects perform an isometric bench-press test involving five maximal isometric contractions before and after 5 d of Cr ingestion at 20 g/d. Five days of Cr supplementation increased total body mass and fat-free body mass, and peak and total force during a repeated maximal isometric bench-press test *(27)*.

Even when provided in lower doses for 3 wk, Cr supplementation can enhance short-duration, high-intensity activities evidenced by the fact that Cr ingested at a dosage of approx 7 g/d combined with resistance training resulted in subjects performing more total work until fatigue, experiencing significantly greater improvements in peak force and peak power, and maintaining elevated mean peak power for a longer period of time. These results indicate that Cr supplementation can significantly improve factors associated with short-duration, high-intensity activity *(28)*.

7.2. Long-Term Performance Effects

Long-term Cr supplementation appears to enhance the quality of training generally leading to 5–15% greater gains in strength and performance. In training studies, subjects taking Cr typically gain twice as much total body mass and/or fat-free mass (an extra 1–2 kg of muscle mass during 4–12 wk of training) than subjects taking a placebo. Most long-term Cr studies have typically involved male participants and observed them to gain approx 4–6 kg of total body mass during 8–12 wk of resistance training.

Rather than use trained athletes, most long-term Cr supplementation studies have involved untrained; therefore, one must use caution when making implications for trained athletes as the physiological responses may likely differ. To identify the effects of 10 wk of Cr supplementation combined with resistance training on muscle performance and muscle morphometric, contractile and molecular characteristics, a study was performed in which untrained male participants received 20 g/d of Cr for the first week of training and 5 g/d for the remaining 9 wk. Cr supplementation resulted in a significant increase in types I and II fiber, a significant increase in calcium sensitivity, and an increase in muscle strength, suggesting that the increase in muscle strength and fiber size to be a result of an increase in calcium influx *(29,30)*.

Cr loading and prolonged supplementation results in increases in muscle-Cr content, body composition, muscle and whole-body oxidative capacity, and substrate utilization during submaximal exercise. A study was conducted to investigate this by using repeated supramaximal bicycle sprints, as well as endurance-type time-trial performance on a cycle ergometer. Participants ingested Cr during a 5-d loading period (20 g/d) after which supplementation was continued for up to 6 wk (2 g/d). Cr loading increased muscle free-Cr, PCr, and total-Cr content; however, the subsequent use of 2 g/d as a maintenance dose resulted in a decline in both the elevated PCr and total-Cr content and maintenance of the free Cr concentration. Both short- and long-term Cr supplementation improved performance during repeated supramaximal sprints on a cycle ergometer. However, whole-body and muscle oxidative capacity, substrate utilization, and time-trial performance were not affected. The increase in body mass following Cr loading was maintained after 6 wk of continued supplementation and accounted for by a corresponding increase in fat-free mass. This study provides definite evidence that prolonged Cr supplementation in humans does not increase muscle or whole-body oxidative capacity and, as such, does not influence substrate utilization or performance during endurance cycling exercise. However, the findings do suggest that prolonged Cr ingestion induces an increase in fat-free mass *(31)*.

Using only a 4-wk protocol, the effects of Cr supplementation was determined on muscle strength in conjunction with resistance training (3 d/wk) in nonresistance-trained males while ingesting 20 g of Cr and 140 g of dextrose each day for the first week, followed by 5 g of Cr and 35 g of dextrose for the remaining 3 wk. Cr supplementation resulted in increases in muscle-Cr uptake, total body mass, and isokinetic and isometric muscle force production, and indicated that 4 wk of Cr supplementation and resistance training can increase total body mass and muscle strength *(32)*.

The effect of 6 wk of Cr supplementation during a periodized program of arm flexor strength training was performed in an effort to determine arm flexor strength, upper arm muscle area, and body composition. Cr monohydrate (20 g/d) were ingested for 5 d, after which Cr supplementation was reduced to 2 g/d for the remainder of the study. Cr supplementation resulted in increases in upper-arm strength and muscle area, total body mass, and fat-free mass, suggesting that Cr supplementation during arm flexor strength training leads to greater increases in arm flexor muscular strength, upper arm muscle area, and fat-free mass than strength training alone *(33)*.

Although most long-term Cr studies have been conducted in males, the effects of oral Cr supplementation on muscle-PCr concentration, muscle strength, and body composition were investigated in young females during 10 wk of resistance training conducted thrice weekly. Four days of Cr ingested at a dosage of 20 g/d increased muscle-PCr concentration by 6%. Thereafter, this increase was maintained during 10 wk of training combined with Cr ingested at a dosage of 5 g/d. Cr supplementation increased maximal muscle strength and fat-free mass and suggested long-term Cr supplementation to enhance the progress of muscle strength during resistance training in females *(34)*.

7.3. Short- and Long-Term Performance Effects in Competitive Athletes

Many studies involving Cr supplementation have used untrained or moderately trained participants who were not trained, competitive athletes. Therefore, because of the inherent differences in muscle metabolism, the responses to Cr supplementation may not be similar to that otherwise observed. The value of short-term Cr supplementation in athletes while involved in training has been documented in a study of male handball players on maximal muscle strength and power production during repetitive high-power-output exercise bouts, repeated running sprints, and endurance; the study was evaluated after ingesting 20 g/d of Cr for 5 d. Cr supplementation significantly increased body mass, number of repetitions performed to fatigue, and total average power output values during the bench press and squat, as well as the vertical jump and running times during sprinting, thereby suggesting that Cr supplementation leads to significant improvements in maximal strength and power *(35)*.

In addition, the effect of oral Cr supplementation on aerobic and anaerobic performance was investigated in elite male rowers during 7 d of endurance training during which time all subjects ingested 20 g of Cr monohydrate for 5 d. Maximal power output did not significantly differ after Cr ingestion; however, the mean lactate threshold rose significantly. In regards to anaerobic endurance capacity, Cr supplementation increased rowing capacity. The results indicate that in elite rowers, Cr supplementation improves endurance and anaerobic performance, independent of the effect of intensive endurance training *(36)*. The effects of Cr loading on the onset of neuromuscular fatigue in female rowers was performed by having subjects ingest 20 g of Cr monohydrate and 80 g of glucose each day for five consecutive days. Results demonstrated that Cr loading was effective in increasing the physical work capacity and suggested that Cr loading may delay the onset of neuromuscular fatigue *(37)*, likely because of augmented levels of PCr in the exercised muscles.

A number of studies have been conducted in competitive athletes. The effects of 4 wk of Cr supplementation during resistance/agility training on body composition, strength, and sprint performance were evaluated in National Collegiate Athletic Association (NCAA) division IA football players. Subjects ingested approx 16 g/d of Cr monohydrate for the entire 28 d. Cr supplementation increase total body mass and fat-free mass, along with gains in bench press lifting volume, the sum of bench press, squat, and power clean lifting volume, and total work performed during 6-s sprints, and suggests Cr supplementation to promote gains in fat-free mass, isotonic lifting volume, and sprint performance during intense resistance/agility training *(38)*.

In another study, NCAA division I football athletes were subjected to 9 wk of Cr monohydrate supplementation coupled with resistance training on body composition and neuromuscular performance. Participants received 20 g/d of Cr for the first 5 d in four 5-g doses followed by 5 g/d for the remainder of the study and resistance trained for 4 d/wk. Cr supplementation was shown to increase total body mass, fat-free mass, intracellular hydration, muscle strength, and anaerobic power capacity, indicating that Cr, supplemented concurrently with resistance and anaerobic training, might positively affect cell hydration status and enhance performance variables compared with resistance training alone *(39)*. In addition, the effect of 50 d of Cr supplementation was determined on collegiate football players during which the subjects continued their normal training schedules. Cr supplementation significantly increased total body mass, fat-free mass, and maximum muscle strength and power *(40)*.

Male and female track and field athletes were provided with 0.3 g Cr/kg body mass/d for 1 wk, followed by 0.03 g Cr/kg body mass/d for 7 wk, and participated in a periodized strength and conditioning program during preseason training. Eight weeks of Cr supplementation resulted in greater increases in total body mass and fat-free mass compared with placebo. Additionally, Cr resulted in significantly greater improvement in initial rate of power production *(41)*.

Sprint cycle and skating performance in ice-hockey players was determined by having participants ingest 20 g/d of Cr monohydrate, after which a maintenance dose of 5 g/d was ingested for 10 wk. Cr supplementation was shown to increase mean power output in addition to on-ice sprint performance thereby demonstrating that Cr supplementation has an ergogenic effect in elite ice-hockey players *(42)*.

8. CONCLUSION

Supplemental Cr monohydrate is an effective nutritional compound in increasing short-term (1–3 wk) and long-term (4–12 wk) muscle performance, as indicated by increase in strength and power. In addition, supplemental Cr appears to also be a beneficial catalyst in increasing muscle hypertrophy and total body mass when combined with resistance training. Cr is typically ingested at an approximate dosage of 20 g/d (0.3 g/kg body mass/d) for 5–7 d during a loading phase and then around 5 g/d (0.07 g/kg body mass/d) for several weeks during a maintenance phase. Evidence suggests that Cr supplementation is more likely suited to improve performance in the sports/activities that require more of an anaerobic performance component in activities such as weight lifting, sprinting, football, and ice hockey. Cr enhances short-term, anaerobic endurance through its ability to enhance muscle bioenergetics. In addition, Cr indirectly promotes muscle growth through long-term supplementation coupled with resistance training by extending exercise output. As a result, muscles then compensate for the increased mechanical load through the production of new muscle proteins. These newly added proteins promote hypertrophy, thereby allowing muscles to generate greater amounts of force and power.

REFERENCES

1. Volek J, Kraemer W. Creatine supplementation: its effect on human muscular performance and body composition. J Strength Cond Res 1996; 10(3):200–210.
2. Prevost M, Nelson A, Morris G. Creatine supplementation enhances intermittent work performance. Res Q Exerc Sport 1997; 68(3):233–240.
3. Greehaff P, Bodin K, Soderlund K, Hultman E. Effect of oral creatine supplementation on skeletal muscle phosphocreatine resynthesis. Am J Physiol 1994; 266 (5 Pt 1):E725–E730.
4. Kreider RD, Willoughby M, Greenwood G, Parise E, Payne M, Tarnopolsky J. Effects of serum creatine supplementation on muscle creatine and phosphagen levels. Exerc Physiol 2003; 6(4):24–33.
5. Kurosawa Y, Hamaoka T, Katsumura T, et al. Creatine supplementation enhances anaerobic ATP synthesis during a single 10 sec maximal handgrip exercise. Mol Cell Biochem 2003; 244(1–2):105–112.
6. Vandenberghe K, Van Hecke P, Van Leemputte M, Vanstapel F, Hespel P. Phosphocreatine resynthesis is not affected by creatine loading. Med Sci Sports Exerc 1999; 31(2):236–242.

7. Mesa J, Ruiz J, Gonzalez-Gross M, Gutierrez Sainz A, Castillo Garzon M. Oral creatine supplementation and skeletal muscle metabolism in physical exercise. Sports Med 2002; 32(14):903–944.
8. Loike J, Zalutsky D, Kaback E, Miranda A, Silverstein S. Extracellular creatine regulates creatine transport in rat and human muscle cells. Proc Natl Acad Sci 1988; 85(3):807–811.
9. Cecil K, Salomons G, Ball W, et al. Irreversible brain creatine deficiency with elevated serum and urine creatine: A creatine transporter defect? Ann Neurol 2001; 49(3):401–404.
10. Guerrero-Ontiveros M, Wallimann T. Creatine supplementation in health and disease. Effects of chronic creatine ingestion in vivo: down-regulation of the expression of creatine transporter isoforms in skeletal muscle. Mol Cell Biochem 1998; 184(1–2):427–437.
11. Walzel B, Speer O, Zanolla E, Eriksson O, Bernardi P, Wallimann T. Novel mitochondrial creatine transporter activity. J Biol Chem 2002; 277(40):37,503–37,511.
12. Snow R, Murphy M. Factors influencing creatine loading into human skeletal muscle. Ex Sport Sci Rev 2003; 31(3):154–158.
13. Odoom J, Kemp G, Radda G. The regulation of total creatine content in a myoblast cell line. Mol Cell Biochem 1996; 158(2):179–188.
14. Nash S, Giros B, Kingsmore S, et al. Cloning, pharmacological characterization, and genomic localization of the human creatine transporter. Receptors Channels 1994; 2(2):165–174.
15. Bessmann S, Geiger P. Transport of energy in muscle: the phosporylcreatine shuttle. Science 1981; 211(4481):448–452.
16. Pearson D, Hamby D, Russel W, Harris T. Long-term effects of creatine monohydrate on strength and power. J Strength Cond Res 1999; 13(3):187–192.
17. Volek J, Duncan N, Mazzetti S, et al. Performance and muscle fiber adaptations to creatine supplementation and heavy resistance training. Med Sci Sports Exerc 1999; 31(8):1147–1156.
18. Willoughby D, Rosene J. Effects of oral creatine and resistance training on myosin heavy chain expression. Med Sci Sports Exerc 2001; 33(10):1674–1681.
19. Willoughby D, Rosene J. Effects of oral creatine and resistance training on myogenic regulatory factor expression. Med Sci Sports Exerc 2003; 35(6): 923–929.
20. Hespel P, Op't Eijnde B, Van Leemputte M, et al. Oral creatine supplementation facilitates the rehabilitation of disuse atrophy and alters the expression of muscle myogenic factors in humans. J Physiol 2001; 536(Pt 2):625–633.
21. Olsen S, Aagaard P, Kadi F, et al. Creatine supplementation augments the increase in satellite cell and myonuclei number in human skeletal muscle induced by strength training. J Physiol 2006; 573(Pt 2):525–534.
22. Volek J, Mazzetti S, Farquhar W, Barnes B, Gomez A, Kraemer W. Physiological responses to short-term exercise in the heat after creatine loading. Med Sci Sports Exerc 2001; 33(7):1101–1108.
23. Powers M, Arnold B, Weltman A, et al. Creatine Supplementation Increases Total Body Water Without Altering Fluid Distribution. J Athl Train 2003; 38(1): 44–50.
24. Haussinger D. The role of cellular hydration in the regulation of cell function. Biochem J 1996; 313(Pt 3):697–710.

25. Tarnopolsky M, Parise G, Yardley N, Ballantyne C, Olatinji S, Phillips S. Creatine dextrose and protein-dextrose induce similar strength gains during training. Int J Sports Nutr Exerc Metab 2000; 10(4):452–463.
26. Okudan N, Gokbel H. The effects of creatine supplementation on performance during the repeated bouts of supramaximal exercise. J Sports Med Phys Fitness 2005; 45(4):507–511.
27. Kilduff L, Vidakovic P, Cooney G, et al. Effects of creatine on isometric bench-press performance in resistance-trained humans. Med Sci Sports Exerc 2002; 34(7):1176–1183.
28. Burke D, Silver S, Holt L, Smith Palmer T, Culligan C, Chilibeck P. The effect of continuous low dose creatine supplementation on force, power, and total work. Int J Sports Med Exerc Metab 2000; 10(3):235–244.
29. Netreba I, Shenkman B, Popov D, et al. Creatine as a metabolic controller of skeletal muscles structure and function in strength exercises in humans. Ross Fiziol Zh Im I M Sechenova 2006; 92(1):113–122.
30. Shenkman B, Litvinova K, Gasnikova N, et al. Creatine as a metabolic controller of skeletal muscles structure and function in strength exercises in humans. The cellular mechanisms. Ross Fiziol Zh Im I M Sechenova 2006; 92(1):100–112.
31. van Loon L, Oosterlaar A, Hartgens F, Heselink M, Snow R, Wagenmakers A. Effects of creatine loading and prolonged creatine supplementation on body composition, fuel selection, sprint and endurance performance in humans. Clin Sci (Lond) 2003; 104(2):153–162.
32. Kilduff L, Pitsiladis Y, Tasker L, et al. Effects of creatine on body composition and strength gains after 4 weeks of resistance training in previously nonresistance-trained humans. Int J Sports Nutr Exerc Metab 2003; 13(4):504–520.
33. Becque M, Lochamann J, Melrose D. Effects of oral creatine supplementation on Muscular strength and body composition. Med Sci Sports Exerc 2000; 32(3):654–658.
34. Vandenberghe K, Goris M, Van Hecke P, Van Leemputte M, Vangerven L, Hespel P. Long-term creatine intake is beneficial to muscle performance during resistance training. J Appl Physiol 1997; 83(6):2055–2063.
35. Izquierdo M, Ibanez J, Gonzalez-Badillo J, Gorostiaga E. Effects of creatine supplementation on muscle power, endurance, and sprint performance. Med Sci Sports Exerc 2002; 34(2):332–343.
36. Chwalbinska-Moneta J. Effect of creatine supplementation on aerobic performance and anaerobic capacity in elite rowers in the course of endurance training. Int J Sports Nutr Exerc Metab 2003; 13(2):173–183.
37. Stout J, Eckerson J, Ebersole K, et al. Effect of creatine loading on neuromuscular fatigue threshold. J Appl Physiol 2000; 88(1):109–112.
38. Kreider R, Ferreria M, Wilson M, et al. Effects of creatine supplementation on body composition, strength, and sprint performance. Med Sci Sports Exerc 1998; 30(1):73–82.
39. Bemben M, Bemben D, Ooftiss D, Knehans A. Creatine supplementation during resistance training in college football athletes. Med Sci Sports Exerc 2001; 33(10):1667–1673.
40. Stone M, Sanborn K, Smith L, et al. Effects of in-season (5 weeks) creatine and pyruvate supplementation on anaerobic performance and body composition in American football players. Int J Sport Nutr 1999; 9(2):146–165.

41. Lehmkuhl M, Malone M, Justice B, et al. The effects of 8 weeks of creatine monohydrate and glutamine supplementation on body composition and performance measures. J Strength Cond Res 2003; 17(3):425–438.
42. Jones A, Atter T, Georg K. Oral creatine supplementation improves multiple sprint performance in elite ice-hockey players. J Sports Med Phys Fitness 1999; 39(3):189–196.

3 Creatine Supplementation in Endurance Sports

Joel T. Cramer, PhD

1. INTRODUCTION

Creatine (Cr) is a compound that is synthesized endogenously in the kidneys, liver, and pancreas by the transamidination and subsequent transmethylation of three constituent amino acids: glycine, agrinine, and methionine *(1)*. As a result of its amino acid origin, Cr can also be manufactured and consumed as a nutritional supplement. Consequently, Cr is currently regarded as a true ergogenic aid, owing to many well-controlled clinical trials that have demonstrated increases in muscle strength, power output, and muscle mass in response to exogenous Cr consumption *(1–6)*. In fact, aside from the potential ergogenic benefits of caffeine, Cr is the most widely marketed nutritional supplement in the world.

Cr was first discovered as an organic constituent of meat by a French scientist, Chevreul, in 1835. Later, Cr was characterized as an essential intermediate in skeletal muscle metabolism in the early 20th century *(7)*. However, because the majority of scientific literature on Cr supplementation as an ergogenic aid has been published since 1992, the beneficial effects of Cr on human performance are relatively new. The paper by Harris et al. *(8)* is perhaps regarded as a landmark study, as it was the first to demonstrate that 20 g/d of oral Cr supplementation for three or more days increased total muscle-Cr stores, which was further augmented by exercise. Since then, a plethora of studies have examined the effects of Cr supplementation on exercise performance tasks that rely heavily on the Cr phosphate (CrP) energy system *(5)*.

From: *Essentials of Creatine in Sports and Health*
Edited by: J. R. Stout, J. Antonio and D. Kalman © Humana Press Inc., Totowa, NJ

It is well known that short-duration (10–20 s) high-intensity (90–100%) exercise relies on ATP resynthesis through the Cr kinase (CK) reaction:

$$CrP + ADP + H^{+} \xleftrightarrow{\text{creatinkinase}} ATP + Cr \qquad (1)$$

The CK reaction resynthesizes ATP from ADP very rapidly, but is limited in capacity, owing to the exhaustible stores of CrP in the muscle. This concept is supported by the findings of Tesch et al. *(9)* who reported that fast-twitch fibers have greater CrP stores than slow-twitch fibers. In addition, fast-twitch fibers had lower CrP concentrations following 30 s of maximal exercise than slow-twitch fibers, which suggested that fast-twitch fibers rely heavily on CrP-mediated ATP resynthesis *(9)*. However, with 5 d of Cr supplementation, Greenhaff and colleagues *(10)* have demonstrated 20–25% increases in total muscle-Cr stores, 20% of which is available as CrP. Therefore, it is not surprising that Cr supplementation augments short-duration high-intensity exercise, which is driven largely by CrP energy system favored by fast-twitch muscle fibers *(11)*. Why, then, would Cr supplementation improve endurance performance, which is most often associated with slow-twitch muscle fibers and their oxidative capacity? The answer to this question may lie in the mechanisms of action of Cr and CrP.

Four primary mechanisms have been suggested to explain the ergogenic benefits of Cr supplementation *(6)*:

1. Increases in total muscle-Cr concentrations will provide more CrP for the CK reaction (Eq. 1). Energy is released by the hydrolysis of ATP to facilitate muscle contraction as a result of the following reaction:

$$ATP + H_2O \xleftrightarrow{\text{ATPase}} ADP + P_i + \text{energy} \qquad (2)$$

During high-intensity exercise, the rapid hydrolysis of ATP will increase the concentration of ADP, which will drive the CK reaction to the right, because of the *Law of Mass Action*. The *law of mass action* or the *mass action effect* is a property of near-equilibrium, reversible reactions that will drive the direction of the reaction based on the concentration of the reactants *(12)*. For example, when the concentration of ADP increases, the CK reaction will proceed to the right to yield more ATP and free-Cr. This is beneficial when a rapid supply of ATP is needed during anaerobic, short-duration high-intensity muscle contractions. Therefore, the higher the concentration of CrP in the muscle, the longer the CK reaction can resynthesize ATP, which translates into an increase in resistance to fatigue and better performance for tasks that rely on the CrP energy system.

2. Not only is the concentration of CrP important, but the free-Cr availability in muscle might also serve an important role in stimulating mitochondrial respiration—a process which has been termed the *phosphorylcreatine (PCr) shuttle*. Originally proposed by Bessman and Geiger *(13)*, the PCr shuttle conceptually involves three sites: (1) the site of energy utilization (i.e., the myosin heads), (2) the site of energy production (i.e., the mitochondria), and (3) the space between the sites of utilization and production (i.e., sarcoplasm). Specifically, CrP resynthesizes ATP at the myosin heads through the CK reaction. The resultant free-Cr travels to the mitochondrial membrane where it is rephosphorylated through the CK reaction (proceeding to the left) by the addition of an ATP molecule synthesized from within the mitochondria. The rephosphorylated CrP then travels back to the myosin heads where it can be used to phosphorylate ADP.

   There are several potential reasons why the PCr shuttle is important for CrP-mediated ATP synthesis *(14)*. First, the PCr shuttle stabilizes the CrP pool in the sarcoplasm, which improves ATP and CrP supply. Second, the constant "shuttling" of phosphates by free-Cr limits the metabolic transformation of Cr to creatinine. Creatinine leaves the muscle cell and is eliminated in the urine by the kidneys. Third, the rephosphorylation of free-Cr at the mitochondria reduces the concentration of free-Cr (particularly during recovery), which maintains a favorable concentration gradient across the muscle cell membrane and facilitates the uptake of Cr from the blood. This might be further enhanced when concentrations of Cr in the blood are elevated during periods of Cr supplementation.

3. The rephosphorylation of ADP through the CK reaction may also act as a proton buffer *(6)*. As depicted in Eq. 1, when the CK reaction proceeds to the right a hydrogen cation (proton, H^+) is utilized, which maintains intracellular pH and helps to prolong fatigue *(15)*. Therefore, in theory, greater concentrations of total Cr might augment this buffering ability. However, studies using ^{31}P magnetic resonance spectroscopy, have demonstrated that Cr supplementation does not alter or maintain higher pH levels during fatiguing contractions in the quadriceps femoris *(16)* or the triceps surae *(17)*.

4. It has also been suggested that CrP can allosterically inhibit the phosphofructokinase (PFK) enzyme, which is the primary rate limiting enzyme of anaerobic glycolysis *(6)*. When CrP concentrations are high, PFK turnover is suppressed. However, as CrP concentrations decrease during high-intensity exercise, PFK may become less inhibited,

which might promote the glycolytic metabolic pathways. With Cr supplementation, CrP concentrations may remain elevated for a longer duration, which may spare the initiation of glycolysis. When ATP demand supersedes ATP supply, there is a muscle energy imbalance. Therefore, supplementing the diet with Cr may suppress the extent of muscle energy imbalance, which has been demonstrated by limited accumulations of inosine monophosphate, ammonia (NH_3), and hypoxanthine during exercise with Cr supplementation *(18,19)*. However, the study by McConnell et al. *(19)* reported no Cr-related alterations in muscle glycogen or lactate concentrations, which should provide insight regarding the relative contributions of the CrP and anaerobic glycolysis energy systems.

1.1. How Can Cr Supplementation Enhance Endurance Performance?

Based on the four proposed mechanisms of action for Cr supplementation above, perhaps the most intriguing are the roles of Cr and CrP in the PCr shuttle for enhancing endurance performance. Indeed, studies have demonstrated that adding Cr to skeletal or cardiac muscle in vitro increases the rate of oxidative phosphorylation *(13,20,21)*, and the turnover rate of mitochondrial CK is directly proportional to the rate of aerobic respiration *(22)*. In addition, it has been suggested that the increased rate of oxidative phosphorylation that occurs when adding Cr to myocytes in vitro is a result of an increased sensitivity of respiration to ADP *(23)*. That is, an increase in ADP concentration is a potent stimulator of cellular respiration and oxidative phosphorylation. The other product of ATP hydrolysis (Eq. 2), inorganic phosphate (P_i), also stimulates the rate of mitochondrial ATP synthesis *(24,25)*. Thus, indirectly, the ratio of CrP to Cr is an important determinant of the concentration of ADP and P_i, such that as the CrP:Cr ratio decreases, ADP and P_i concentrations increase (*see* Eqs. 1 and 2). Therefore, as Cr supplementation decreases the CrP:Cr ratio, because of the disproportionate increase in free-Cr compared with CrP *(8)*, it is possible that Cr loading might stimulate oxidative phosphorylation *(26)*.

In addition, slow-twitch muscle fibers might be more sensitive to Cr-related increases in respiration rates *(23)* than fast-twitch fibers *(27)*. In fact, Tonkonogi et al. *(28)* have demonstrated that the number of slow-twitch fibers is directly related to the rate of aerobic respiration when Cr is added to a culture of human skeletal muscle fibers. This suggested that individuals with a greater proportion of slow-twitch fibers, such as endurance athletes, might experience greater levels of oxidative phosphorylation after Cr supplementation. An alternative hypothesis is that increases in intramuscular CrP might allow calcium adenosine triphosphatase (an enzyme that catalyzes the hydrolysis of ATP to pump calcium back into the sarcoplasmic reticulum during muscle relaxation) to operate at a higher thermodynamic efficiency and thereby decrease the energy cost of muscle relaxation *(29,30)*. However, these mechanistic hypotheses are largely based on basic, physiological, in vitro experiments. To fully understand how Cr supplementation may or may not impact endurance exercise performance, measures of endurance must be assessed with and without Cr supplementation and these findings must be synthesized and evaluated objectively. Therefore, this chapter reviews the evidence on whether Cr supplementation affects endurance performance.

2. METHODS

With the push toward evidence-based recommendations, many practitioners in coaching, strength and conditioning, personal training, dietetics, rehabilitation, and athletic training are now focusing on whether Cr is "significantly" (traditionally defined as a type I error rate of 5% or less; $p \leq 0.05$) superior to a placebo (Pl) or control group, based on only one or two studies. As an alternative, there are many narrative literature reviews that provide interpretations of several representative articles *(1–4)*, but these types of reviews can be subjective. As few practitioners have the time or resources to conduct an exhaustive review of the literature, *systematic reviews* or *meta-analyses (5)* have become important tools that practitioners can use to make evidence-based recommendations. Because the purpose of this chapter is to provide an objective literature review, this section will outline the methods that were used to choose the articles to be reviewed and how the aggregate data from these articles were compared with formulate conclusions.

PubMed (www.pubmed.gov, a service of the United States National Library of Medicine and the National Institutes of Health) was searched for all articles related to Cr supplementation and endurance performance. The primary search terms were *creatine* and *supplementation* and *endurance* and *performance*, and the search was delimited to studies conducted on humans and articles written in the English language. The types of studies included in the search based on the PubMed search limitations included clinical trials; randomized control trials; classical articles; clinical trial phases I, II, III, or IV; and controlled clinical trials. The search conducted in August of 2006 with the aforementioned delimitations yielded 43 articles. Each article was scanned for content, and the primary qualifications for inclusion were that the study examined some measure of muscle or whole-body endurance and Cr supplementation by itself was a unique condition. As a result, 18 of the 43 articles were reviewed from the PubMed search. In addition, all pertinent articles from the bibliographies of these studies were included, which brought the total number of articles reviewed to 35.

The results in Table 1 were presented as they were in the original articles. To compare results across studies in separate figures, the *relative effects* (RE) were calculated for each study similar to the methods of Shrier *(31)*. The mean values for each study were either reported in the text, tables, or graphs. When reported in graphical format, the values were extrapolated with a ruler to the *y*-axis and estimated by position.

For mixed factorial designs, the RE were calculated with the following equation, which was a modified version of Shrier's *(31)* equation to account for the time- (pre vs post) and group-dependent (Cr vs Pl) changes:

$$\mathrm{RE} = \left[\frac{\left(\dfrac{\mathrm{Post}_{\mathrm{Cr}}}{\mathrm{Pre}_{\mathrm{Cr}}}\right)\times 100}{\left(\dfrac{\mathrm{Post}_{\mathrm{Pl}}}{\mathrm{Pre}_{\mathrm{Pl}}}\right)\times 100}\right]\times 100 \qquad (3)$$

where $\mathrm{Pre}_{\mathrm{Cr}}$ is the pretesting measure of endurance for the Cr group, $\mathrm{Post}_{\mathrm{Cr}}$ is the posttesting measure of endurance for the Cr group, $\mathrm{Pre}_{\mathrm{Pl}}$ is the pretesting measure of endurance for the Pl group, and $\mathrm{Post}_{\mathrm{Pl}}$ is the post-testing measure of endurance for the Pl group. This equation was used to account for the changes from pre- to post-testing in the Cr and Pl groups, which was the most common research design. For simple between-group comparisons, the RE was calculated with the following equation *(31)*:

$$\mathrm{RE} = \left(\frac{\mathrm{Endurance}_{\mathrm{Cr}}}{\mathrm{Endurance}_{\mathrm{Pl}}}\right)\times 100 \qquad (4)$$

where $endurance_{\mathrm{Cr}}$ is the measure of endurance for the Cr group, and $endurance_{\mathrm{Pl}}$ is the endurance measurement for the Pl group. For simple repeated measures designs, the RE was calculated with the following equation *(31)*:

$$\mathrm{RE} = \left(\frac{\mathrm{Post}_{\mathrm{Cr}}}{\mathrm{Pre}_{\mathrm{Cr}}}\right)\times 100 \qquad (5)$$

It should be noted that all RE estimates were calculated with mean values reported in the articles, which is not equivalent to the true RE (i.e., the RE should be calculated for each individual, and then the mean

Table 1

Detailed Summary of Studies Examining Whether Cr Supplementation Improved Endurance Performance

Population	*Design*	*Intervention*	*Results*	*Comments*	*References*
Habitually active to well-trained men; 19–37 yr; $n = 18$	Nonrandomized, double-blinded, Pl-controlled, mixed factorial design; two groups: (1) Cr ($n = 9$) and (2) Pl ($n = 9$); testing occurred before and after a 6-d supplementation period	Cr group received 5 g of Cr + 1 g of glucose, whereas the Pl group received 6 g of glucose per day for 6 d; incremental treadmill running tests, supramaximal treadmill running tests, and 6 km terrain running tests were performed before and after the supplementation; performance time, VO_2 (30 s, 60 s, and peak), heart rate, blood lactate, and plasma hypoxanthine were measured	Cr decreased the terrain running performance and increased blood lactate after the supramaximal treadmill test. No other changes were observed for either the Cr or Pl groups	Cr did not improve endurance times, VO_2 heart rate, blood lactate, or plasma hypoxanthine	*36*

(Continued)

Table 1 *(Continued)*

Population	*Design*	*Intervention*	*Results*	*Comments*	*References*
Recreationally active men; age ± SD = 21 ± 1 yr; $n = 17$	Randomized, double-blinded, Pl-controlled, mixed factorial design; two groups: (1) Pl ($n = 8$) and (2) Cr ($n = 9$). Tests were administered before and after a 4-d Pl administration to all subjects, and before and after another 4-d supplement period where half took Cr and half took the Pl	Pl doses were 4 × 10 g/d of glucose, whereas Cr doses were 4 × 70 mg of Cr + 5 g of glucose per day for 4 d. Incremental tests for VO_{2peak} and seven 10-s cycling sprints were performed before and after each supplementation period. Peak and mean power output, plasma lactate, pH, excess postexercise oxygen consumption (EPOC), and VO_{2peak} were measured during the testing sessions	Cr had no effect on any of the dependent variables	Cr did not improve VO_{2peak} or multiple repeated cycling sprint trials	*37*
Well-trained male endurance cyclists (age ± SD = 24 ± 1 yr;	Randomized, double-blinded, Pl-controlled, mixed factorial design; two groups: (1) Cr ($n = 10$) and (2) Pl	Cr group consumed 20 g/d of Cr, whereas the Pl group consumed 20 g/d of a Pl for 7 d; 1-h time trials on competition bicycles	Cr had no impact on the performance of the cycling time trials (distance covered or heart	Despite no changes in exercise performance, the authors suggested that	*18*

$n = 20$)	($n = 10$); testing occurred before and after a 7-d supplementation period	were performed and heart rate was monitored; muscle biopsies were performed before each cycling test; blood samples were taken throughout the cycling test; total Cr, lactate, ammonia, hypoxanthine, and urate concentrations were determined	rate); lactate concentrations were unaltered by Cr; however, plasma ammonia were lower as a result of Cr	Cr may reduce adenine nucleotide degradation and alter metabolism during endurance exercise	
Healthy, nonathletic elderly women; 60–80 yr; $n = 16$	Randomized, mixed factorial, single-blinded, Pl-controlled design; testing was completed before and after a 7-d supplementation period	Seven days of supplementation: Cr ($n = 10$) or cellulose Pl ($n = 6$) at 0.3 g/kg body mass. Blood analyses, incremental cycling tests, constant-load cycling tests, and functional tests were performed. Standard blood chemistry; VO_{2max},	Cr improved sit-to-stand times; however, no other changes were observed for any other variables. Cr did not alter body mass	Cr did not affect any long-duration markers of endurance performance or functional capacity, but Cr did improve short-duration lower body functional tasks.	*38*

(Continued)

Table 1 *(Continued)*

Population	*Design*	*Intervention*	*Results*	*Comments*	*References*
		respiratory exchange ratio (RER), power output (Wmax), and VT were assessed during the incremental test, whereas gross mechanical efficiency (GE), blood lactate (Bla), and ammonia (NH_3) were assessed during the constant-load test. One-mile walk tests and sit-to-stand tests were performed for functional tests		These findings might be related to the age-dependent changes in muscle morphology	
Male elite rowers; 20–31; $n = 16$	Randomized mixed factrorial, double-blinded, Pl-controlled, design; testing was completed before and after the supplementation period	Seven days of endurance training + 5-d supplement protocol, Cr-group ($n = 8$) ingested 20 g/d of Cr and the Pl-group ($n = 8$) ingested 20 g/d of glucose; incremental intermittent rowing and	No Cr-related changes for heart rate, LAT-4 m*M*, or the patterns of lactate accumulation during of following the incremental test. Work load (W)	Cr improved the LT, anaerobic capacity, and possibly maximal power output during rowing ergometry	*56*

		all-out aerobic rowing tests were performed on ergometers; multiple blood samples were taken. Heart rate, workload at the onset of blood lactate (LAT-4 m*M*; 4 mmol/L); workload at the LT (LAT-log); TTE during the anaerobic test	and blood lactate (mmol/L) at the blood LT increased for the Cr-group, and the TTE for the anaerobic test increased for the Cr-group. A trend toward higher maximal workloads during the incremental test was also noted for the Cr-group		
Well-trained men who participated in resistance training, running, and high-intensity interval training; $n = 11$	Randomized, double-blinded, Pl-controlled, mixed factorial design; two groups: (1) Pl and (2) Cr. Tests were administered before and after the 10-d supplement period	Cr group received 5 g of Cr + 1 g of glucose, whereas the Pl group received 6 g of glucose four times per day for 4 d and twice per day for 6 d; treadmill running tests were performed for TTE and blood lactate was measured	Cr supplementation resulted in nonsignificant improvements in TTE, where the second run were more improved than the first run; blood lactate was higher after Cr supplementation	Cr showed trends toward improvement in running TTE with an emphasis on improvements for the repeated bout effect	*60*

(*Continued*)

Table 1 *(Continued)*

Population	*Design*	*Intervention*	*Results*	*Comments*	*References*
Moderately to highly physically active women; 19–34 yr; $n = 10$	Randomized, double-blinded, Pl-controlled, repeated measures crossover design; two groups: (1) Pl and (2) Cr; groups were switched after 5 wk washout period; testing occurred at baseline and after 2 and 5 d of supplementation	Each dose contained either 18 g of dextrose (Pl) or 5 g of Cr + 18 g dextrose (Cr); the CP test was performed on a cycle ergometer to determine the AWC and body mass was monitored	Cr improved AWC after 5 d of supplementation, but not after 2 d. Body mass increased for both the Cr and Pl groups from baseline to 2 d of supplementation	Cr loading for 5 d resulted in a 22% increase in AWC for women	*48*
Moderately to highly physically active men ($n = 31$, 19–26 yr) and women ($n = 30$, 18–35 yr)	Randomized, double-blinded, Pl-controlled, mixed factorial design; three groups: (1) Pl, (2) Cr, and (3) Cr-phosphate; testing was performed at baseline and after 2 and 6 d of supplementation	Each dose contained either 18 g of dextrose (Pl), 5 g of Cr citrate + 18 g dextrose (Cr), or 2.06 g of sodium phosphate + 2.06 g of potassium phosphate + 5 g of Cr citrate + 18 g dextrose (Cr-phosphate); the CP test was performed on a	Cr-phosphate improved absolute AWC in the men, but not the women; Cr alone did not improve absolute AWC in men or women; AWC relative to body mass was improved	Cr-phosphate may be better than Cr alone for improving AWC in men	*47*

		cycle ergometer to determine the AWC and body mass was monitored	by Cr-phosphate and Cr supplementation for the men, but not the women; body mass increased with supplementation for both men and women		
Healthy, recreationally active, untrained men; 55–75 yr; $n = 46$	Match paired, mixed factorial, double-blinded, Pl-controlled design; testing occurred at baseline, 3, 6, and 12 mo during the 1-yr training and supplementation period	Cr group ($n = 23$) received 5 g/d of Cr tablets, whereas the others ($n = 23$) received Pl tablets; assessments included maximal and submaximal VO_2 RER, V_{slope} isometric peak torque, isokinetic peak torque, and leg extension endurance; body composition; muscle biochemistry and histochemistry; and blood chemistry	No Cr-related changes in isometric or isokinetic strength or endurance, body composition, or submaximal or maximal cycle ergometry endurance performance	Long-term Cr supplementation does not improve physical fitness in older men (55–75 yr)	*39*

(Continued)

Table 1 *(Continued)*

Population	*Design*	*Intervention*	*Results*	*Comments*	*References*
Regional class, competitive triathletes; 22–27 yr; $n = 12$	Nonrandomized, nonblinded, repeated measures design without a Pl; all 12 subjects performed the testing before and after the 5-d supplementation period	Each athlete consumed 6 g/d of Cr for 5 d; a special cycle ergometer test was designed to measure constant workload endurance performance before and after 20, 15-s interval cycle ergometry bouts at high workloads with 45-s rest intervals. Cr, creatinine, lactate, glucose, heart rate, and VO_2 were measured at various intervals during the test	Cr supplementation increased Cr and creatinine concentrations and improved interval cycle ergometry power performance. Cr had no effect on the constant workload endurance performance, heart rate, VO_2, CK, or lactate concentrations	Cr may improve high-intensity interval cycle ergometry performed during an endurance exercise event, but does not improve the overall endurance performance	*14*
Experienced, well-trained Spanish male handball	Randomized, double-blinded, Pl-controlled, match paired mixed factorial design; two groups: Cr group	Cr group consumed 20 g/d of Cr, whereas the Pl group consumed 20 g/d of maltodextrine for 5 d; testing was conducted	Cr increased body mass and improved the number of repetitions to fatigue, average	Cr improved exercise performance that relied primarily on the	*57*

players; $n = 19$	($n = 9$) and Pl group ($n = 10$); the 2-d testing protocol was completed before and after 5-d supplementation period	to examine maximal strength, muscle power output during repetitive bouts, repeated sprint performance; and endurance; exercises included 1 repetition maximum (1RM) and multiple repetition tests for the back squat and bench press, counter movement jumps, repeated intermittent 15-m sprinting, and a multistage discontinuous incremental running test for endurance	power output, lower body 1RM strength, jump height, and sprint times. No changes were seen in the Cr group for upper-body maximal strength or endurance performance	phosphagen system, but not endurance performance	
Moderately active young men who were unversity students; 19–28 yr; $n = 14$	Randomized, double-blinded, Pl-controlled, mixed factorial design; Cr group ($n = 7$) and Pl group ($n = 7$) performed the testing before	Cr group ingested 20 g of Cr + 20 g of maltodextrin per day for 5 d, whereas the Pl group consumed 20 g of maltodextrin per day for 5 d, assessments	Cr reduced body mass, but the Pl did not, Cr did not significantly affect any of the voluntary or stimulated muscle	Short-term Cr supplementation did not improve isometric forearm flexion strength, percent voluntary	*61*

(*Continued*)

Table 1 *(Continued)*

Population	*Design*	*Intervention*	*Results*	*Comments*	*References*
	and after a 5-d supplementation period	were conducted for isometric maximal voluntary contraction (MVC) strength, muscle activation, electrically stimulated contractile properties, surface electromyography, TTE during a fatiguing task, and recovery from fatigue	contractile properties or resistance to fatigue	activation, stimulated contratile properties, TTE, or recovery from fatigue	
Recreationally active but not highly trained participants; age ± SD = 28 ± 3 yr; $n = 9$ (two women, seven men)	Randomized, mixed factorial, crossover design, with group 1 ($n = 5$) performing the control condition first followed by the Cr (Cr) loaded condition; group 2 ($n = 4$) performed the Cr loaded condition first followed by the	Cr condition involved 20 g/d of Cr monohydrate for the first 5 d followed by 5 g/d for the remaining days of the exercise protocol. Subjects completed multiple 6-min cycle ergometer bouts at two workloads: (1) 80% of the VT and	Cr reduced VO_2 and blood lactate during the high-intensity workload (50% between VT and VO_{2max}), which was correlated ($r = 0.87$) with the percentage of fast-twitch fibers	Cr may increase the efficiency of submaximal cycle ergometry at workloads above VT, due to the reduction in VO_2 at the high intensity workload that was related to the	*26*

	control condition 35–50 d later	(2) half the distance between VT and VO_{2max}. O_2 uptake kinetics were assessed as well as blood lactate and end-exercise heart rate during the Cr and control conditions. Biopsies were taken from the vastus lateralis	in the vastus lateralis. Cr did not alter any other variable during either the high- or low-intensity workloads	percentage of type II fibres. CR may reduce the volume of muscle activated at submaximal workloads	
Healthy men that were engaged in a weight training program; age ± SD = 28 ± 4 yr; *n* = 10	Randomized, double-blinded, Pl-controlled, mixed factorial with a crossover design; the Cr group (Cr) and Pl group (Pl) experience both conditions after a 3-d washout period	Each participant was assigned to either group A or B and consumed 10 g of Cr + 92.5 g of glucose per day for 5 d (days 2–6) then consumed only 98.2 g of glucose per day for 5 d (days 9–13) in random order; tests were conducted at baseline, after the first supplementation period (day 8), and after the second supplementation	Cr increase body weight, urinary Cr excretion, MVC strength, and TTE during the leg extensor fatigue trials when compared with Pl	Increase in MVC strength and endurance capacity (<170 s) may have been due to Cr-induced hypertrophy	*62*

(Continued)

Table 1 *(Continued)*

Population	*Design*	*Intervention*	*Results*	*Comments*	*References*
		period (day 15); maximal voluntary contraction (MVC) strength of the leg extensors and TTE at 80, 60, 40, and 20% of MVC were determined			
Competitive, endurance-trained male cyclists or triathletes; age ± SE = 21 ± 1 yr; *n* = 7	Nonrandomized, repeated measures, crossover design, with the control (CON) trial first, and the Cr (CREAT) trial second, separated by 1 wk	42 g/d of dextrose for 5 d before CON trial; 21 g/d of dextrose + 21 g/d of Cr monohydrate for 5 d before CREAT trial. 45-min cycle ergometry at 78% VO_{2peak} followed by a timed ride to complete 250 KJ of work (termed *performance ride*). Assessments were (1) muscle glycogen, (2) lactate, (3) adenosine	Muscle-Cr was increased during the CREAT trial. IMP was the same at rest and after the 45-min ride, but was lower for the CREAT trial after the performance ride. No other meaningful differences between CREAT and CON were observed	IMP reflects muscle energy imbalance (i.e., ATP demand exceeds ATP supply), and may be more sensitive than traditional measures (i.e., ammonia, hyposanthine). Since IMP was attenuated by CREAT, but performance	*19*

		triphosphate (ATP), (4) adenosine diphosphate (ADP), (5) adenosine monophosphate (AMP), (6) inosine monophosphate (IMP), (7) ammonia, (8) CrP, (9) Cr, and (10) hypoxanthine from muscle and/or blood samples. Oxygen uptake, respiratory exchange ratio, and performance time		was unaltered, the authors concluded that performance may be dictated by factors unrelated to muscle energy imbalance	
Elite male surf ski or white-water kayak paddlers; age ± SD = 21 ± 5 yr; $n = 16$	Randomized, nonblinded repeated measures, crossover design; pre- and postsupplementation tests were performed during a Cr and Pl condition, which was separated by a 4-wk washout period	20 g/d of Cr was administered for 5 d for the Cr condition, whereas 20 g/d was administered for the Pl condition; time trials were performed on a wind-braked kayak ergometer for 90, 150, and 300 s durations in random order; work	Cr increased body mass and the work completed during all three time trials. Cr had no affect on peak power. Cr also had increased blood lactate concentrations	Cr increased the work acco-mplished, but not the peak power, during the kayak ergometry time trials between 90 and 300 s	*55*

(Continued)

Table 1 *(Continued)*

Population	*Design*	*Intervention*	*Results*	*Comments*	*References*
		completed (kJ) and peak power (W) were assessed during the time trials	for the 150- and 300-s time trials		
Healthy men; age ± SD = 20 ± 1 yr; $n = 8$	Randomized, double-blinded Pl-controlled, crossover design; postsupplementation testing was conducted after Cr and Pl conditions, which was separated by a 6-wk washout period	20 g/d of Cr was administered for 5 d for the Cr condition, whereas 20 g/d of glucose was administered for the Pl condition; each subject performed four to five cycle ergometry bouts at workloads between 175 and 400 W that lasted 2–10 min after the Cr and Pl conditions. CP and AWC were determined based on the workload-duration curves. VO_2, V_E, VCO_2, heart rate, and maximal	Cr increased the AWC, but not the CP, determined by the workload-duration curves. Cr also minimized the percent decline in leg extension strength. However, Cr did not affect VO_2, V_E, VCO_2, or heart rate	Cr content seems to be an important determinant of AWC, but not CP, during cycle ergometry	*40*

		isometric leg extension strength and endurance was measured			
Men who were participating in university team sports; age ± SD = 24 ± 3 yr; $n = 18$	Randomized, mixed factorial, double-blinded, Pl-controlled design; pre-, mid-(after the 7-d loading phase), and posttests were conducted after a 28-d supplementation period	Cr or Pl groups ingested 20 g/d for the first 7 d, followed by 21 d of 10 g/d; Pl was microcystalline cellulose; assessments were for cardiac structure and function and aerobic power. Heart rate, submaximal and maximal Vo_2, blood pressure, and TTE was measured during incremental cycle ergometer tests. Doppler echocardiographic examinations were performed for cardiac structure and function	With Cr, body mass increased whereas maximal heart rate and submaximal VO_2 at 75 and 150 W decreased. NO other meaningful changes were observed in the Cr or Pl groups	Cr did not affect cardiac structure or function, but it did not improve cycling efficiency at submaximal workloads	*41*
Healthy, physically	Randomized, double-blinded, Pl-controlled,	Cr groups received 20 g/d of Cr capsules; Pl	Cr increased the TTE and VT and	Cr may allow greater cycling	35

(Continued)

Table 1 *(Continued)*

Population	*Design*	*Intervention*	*Results*	*Comments*	*References*
active university students; 20 men (age ± SD = 25 ± 2 yr) and 16 women (24 ± 2 yr)	mixed factorial design; four groups: (1) Cr-men, (2) Cr-women, (3) Pl-men, and (4) pl-women; testing occurred before and after the 7-d supplementation period	groups received 20 g/d of sugar for 7 d; graded exercise tests were performed on a cycle ergometer before and after the supplementation; VO_{2peak}, VO_2 at each stage, VT, total work time, and heart rate were measured	decreased VO_2 and heart rate during the initial stages of the graded exercise test. Cr had no affect on VO_{2peak}	efficiency at submaximal workloads and prolong fatigue, but may not alter maximal oxygen consumption rates	
Healthy, young, competitive male soccer players; age ± SD = 17 ± 2 yr; *n* = 20	Randomized, nonblinded, mixed factorial design; Cr group (*n* = 10) and Pl group (*n* = 10): were matched for age, mass, height, and VO_{2max}, testing occurred before and after a 7-d supple-mentation protocol	Cr group consumed 3 × 10-g doses of Cr over 7 d; Pl took cellulose pills; pre- and postsupplementation soccer-specific skills were accessed, including (1) dribble test, (2) sprint power test, (3) vertical jump test, and (4) endurance test	Cr improved dribble test times, sprint-power performance, and vertical jump height. Cr had no effect on endurance time	Cr improved short-duration, soccer-specific skills, but not endurance	*63*

Physically active, healthy men (age ± SD = 24 ± 1 yr; $n = 10$) and women (24 ± 1 yr; $n = 8$)	Nonrandomized, double-blinded, Pl-controlled, repeated measures crossover design; four groups: (1) Cr-men, (2) Cr-women, (3) Pl-men, and (4) Pl-women; testing the 11-d supplementation periods separated by a 10-d washout period	Cr groups received 18.75 g/d of Cr capsules for 5 d followed by 2.25 g/d for 6 d; Pl groups received 5 g/d of calcium chloride for 5 d followed by 0.6 g/d for 6 d; graded exercise tests were performed initially to determine peak workload; subsequently, TTE was measured with workloads of 150% of peak during (A) continuous riding, (B) intermittent 30-s cycling + 60-s rest, (C) intermittent 20-s cycling + 40-s rest, and (D) intermittent 10-s cycling + 20-s rest; blood samples were collected and analyzed for plasma lactate	Cr increased the TTE for all workbouts (A, B, C, and D); however, the shorter the intermittent work periods, the larger the benefit of Cr supplementation was noticed	Cr may improve the capacity to maintain a specific level of high-intensity, intermittent exercise	*54*

(Continued)

Table 1 *(Continued)*

Population	*Design*	*Intervention*	*Results*	*Comments*	*References*
Healthy, physically active, untrained, nonvegetarian men ($n = 11$) and women ($n = 6$); 18–27 yr	Nonrandomized, nonblinded, Pl-controlled, mixed factorial design, CREAT group ($n = 9$) vs CON ($n = 8$); testing occurred before and after 28 d of Cr supplementation and endurance training	CREAT and CON groups ingested 20 g/d for the first 7 d, followed by 21 d of 5 g/d; Pl was maltodextrin powder; subjects trained on cycle ergonometers for 45 min, 3 d/wk for 4 wk; assessments were VO_2, RER, carbohydrate oxidation, total work, mean power, body consumption, and muscle glycogen content	There were training-induced increases in muscle glycogen, total work, mean power, and carbohydrate oxidation rates; however, these findings were independent of Cr supplementation. There were no changes in body composition or VO_2	Four weeks of relatively low-intensity endurance training elicited positive training adaptations, but these were not enhanced with Cr supplementation	*42*
Well-trained, elite competitive male cyclists;	Match paired (VO_{2max}) and randomized, mixed factorial, double-blinded, Pl-controlled design;	Cr group took 20 g/d of Cr, whereas Pl group took 20 g/d of lactose for 5 d; pre- and postsupplementation	O_2 consumption during the 1st and 2nd 90% bouts was higher with Cr, NH_3, was	Cr improved alternating-intensity cycle ergometry, which may have	*44*

age ± SD = 23 ± 1 yr (Cr group, *n* = 7) and 25 yr ± 2 yr (Pl group, *n* = 7)	Cr and Pl groups were tested before and after a 5-d loading period	tests were conducted for O_2 consumption, TTE, and blood metabolites (lactate, NH_3, and plasma uric acid) during a cycle ergometer test with 3-min bouts alternating between 30% and 90% of their maximal power output	lower during the 1st 90% bout and 3rd 30% bout with Cr; uric acid was lower at exhaustion and 5-min post; TTE increased with Cr	been because of augmented oxidative phosphorylation	
Competitive, club standard oarsmen (age ± SD = 23 ± 4 yr; *n* = 28) and oarswomen (23 ± 5 yr; *n* = 10)	Double-blinded, Pl-controlled, mixed factorial design; two groups were match-paired for gender, body mass, and 1000-m rowing performance into a Cr and Pl group; testing occurred before and after a 5-d supplement period	Participants received 0.25 g/kg body mass of Cr per day for 5 d in the Cr group, whereas the Pl group consumed the same drink without Cr, 1000-m rowing trials were performed before and after the supplementation period and TTE was recorded; Cr uptake was also estimated based on urinary Cr excretion	Cr improved 1000-m rowing times by an average of 2.3 s, and whole body Cr stores increased as a result of the 5-d Cr loading	Cr supplementation improved maximal 1000-m rowing performance	*64*

(Continued)

Table 1 *(Continued)*

Population	*Design*	*Intervention*	*Results*	*Comments*	*References*
Healthy, untrained but active university students; eight men (age ± SD = 23 ± 4 yr) and seven women (23 ± 3 yr)	Randomized, double-blinded, Pl-controlled, mixed factorial design; two groups: (1) Pl and (2) Cr. Tests were administered before and after the 5-d supplement period	Pl group ingested 4 × 6 g of glucose per day; Cr group ingested 4 × 5 g of Cr per day for 5 d; four separate bouts of cycle ergometry exercise were performed at different workloads selected to elicit fatigue within 90–600 s. CP and AWC were determined based on the workload-duration curves. TTE was recorded for each bout	Cr improved AWC, but not CP. Cr also improved the TTE for the two highest workloads	Cr supplementation improved AWC but not Cp, and increased the TTE for the shorter, higher-intensity workloads	*65*
Healthy, physically active, nonmedicated subjects; three men	Nonrandomized, single-blinded, Pl-controlled, crossover design; testing occurred after the 5-d Pl period, then 7–14 d later	Pl condition involved 20 g/d of granulated sugar; Cr condition involved 20 g/d of Cr; isometric and dynamic leg extension exercises	Cr elicited increases in resting PCr, which declined during exercise. However, the isometric and	Cr did not affect the muscle ATP cost of contraction, strength, or contraction,	*16*

(age ± SD = 33 ± 4 yr) and six women (28 ± 3 yr)	after the 5-d Cr supplementation	were performed in a 1.5-TMRI system after each condition; energy costs of the leg extension contractions were determined using P-magnetic resonance spectroscopy (P-MRS)	dynamic costs of contraction, ATP, Pl, pH, PCr resynthesis rate, muscle strength, and endurance were unaffected by Cr supplementation	strength, or endurance during leg extension exercises	
Healthy men; age ± SD = 20 ± 2 yr; *n* = 26	Randomized, double-blinded, Pl-controlled, mixed factorial design; three groups: (1) Pl, (2) Cr, and (3) Cr-phosphate; testing was performed at baseline and after 6 d of supplementation	Each dose contained either 35 g of dextrose (Pl), 5.25 g of Cr + 1 g dextrose (Cr), or 633 mg of sodium and potassium phosphates + 5.25 g of Cr + 33 g dextrose (Cr-carbohydrate); the CP test was performed in a cycle ergometer to determine the AWC and body mass was monitored	Cr and Cr-carbohydrate improved AWC after 6 d of loading; increase in body mass were not significant	Cr and Cr-carbohydrate increased the AWC by 9% and 31%, respectively, which suggested that when carbohydrates + Cr may augment performance beyond Cr alone	*49*
Female competitive	Randomized, mixed factorial, double-	Five days of supplementation: Cr	BW and PWC_{FT} increased as a	PWC_{FT} is a valid, reliable, and	*50*

(Continued)

Table 1 *(Continued)*

Population	*Design*	*Intervention*	*Results*	*Comments*	*References*
university crew team members; age ± SD = 19 ± 2 yr; $n = 15$	blinded, Pl-controlled design; testing was completed before and after a 5-d supplementation period	(20 g Cr + 80 g dextrose per day; $n = 7$) or Pl (80 g dextrose per day; $n = 8$). Body mass (BM) and the physical working capacity at fatigue threshold (PWC_{FT}) were measured before and after the supplementation	result of Cr consumption	sensitive index of neuro-muscular fatigue. Therefore Cr may delay the onset of neuromuscular fatigue in trained women	
Men (age ± SD = 25 ± 5 yr; $n = 51$)	Randomized, double-blinded, Pl-controlled, mixed factorial design; four groups: (1) Pl group and (2) Cr group, (3) β-alanine (β-Ala), and (4) Cr + β-Ala (CrBA); testing occurred before and	All supplements were administered four times per day for the first 6 d then twice per day for 22 d. individual doses were 34 g dextrose (Pl, $n = 13$), 5.25 g Cr + 34 g dextrose (Cr, $n = 12$), 1.6 g β-Ala + 34 g dextrose	CrBa and β-Ala improved the PWC_{FT} but the Pl and Cr had no effect	β-Ala supplementation may delay the onset of neuromuscular fatigue, but there appeared to be no unique or additive benefits of Cr supplementation	*51*

	after the 28-d supplementation period	(β-Ala, $n = 12$), and 5.25 g Cr + 1.6 g β-Ala + 34 g dextrose (CrBA, $n = 14$) The physical working capacity at fatigue threshold (PWC_{FT}) was used as a measure of neuromuscular fatigue before and after the supplementation period			
Physically active men; age ± SD = 26 ± 3 yr; $n = 8$	Nonrandomized, nonblinded, repeated measures design without a Pl; all eight subjects performed the testing before and after the 5-d supplementation period	Each subject consumed 20 g/d of Cr for 5 d; a continuous incremental treadmill running test at workloads of 50, 60, 65, 70, 75, 80, and 90% of VO_{2max} was performed before and after the 5-d supplement period. VO_2, RER, and blood lactate (Bla) were measured at each stage and every 5 min during recovery	Cr had no effect on VO_2, RER, or Bla	Cr supplementation did not influence substrate utilization during endurance excercise to exhaustion	*45*

(Continued)

Table 1 *(Continued)*

Population	*Design*	*Intervention*	*Results*	*Comments*	*References*
Healthy, male ($n = 12$) and female ($n = 11$) competitive, university rowers; mean age = 23 yr	Randomized, double-blinded, Pl-controlled, mixed factorial design; Cr group ($n = 11$) and Pl group ($n = 12$) were matched for gender and rowing performance; both groups performed a training regiment of resistance training and high-intensity rowing; tests were performed at baseline, after loading, after 6 wk of supplementation	For Cr loading, 0.3 g/kg/d of Cr was taken for 5 d, and 0.03 g/kg/d was taken for the remaining 5-wk; Pl was an isocaloric flavored drink without Cr; anthropometric assessments of body composition, total body water, VO_{2max}, VT, 2000-m rowing performance, six repeated intermittent 250-m rowing sprints, and 10RM leg press and bench press with repeated sets were determined at baseline, after loading, and after 6 wk	Cr loading and maintenance increased urinary Cr and creatinine, but did not augment any performance measure beyond training + Pl	No ergogenic benefits of Cr were observed in this study	*66*

Nonvegetarian female athletes from a university swim team; $n = 10$	Randomized, Pl-controlled, nonblinded, mixed factorial design; two groups: (1) Pl and (2) Cr, testing occurred before and after a 6-wk supplementation period	Subjects received either 2 g/d of Cr or 2 g/d of Pl for 6 wk; P magnetic resonance spectroscopy (PMRS) and near-infrared spectroscopy (NIRS) were used to measure the constituents of the CK reaction and muscle reoxygenation, respectively, during exhaustive plantarflexion contractions and finger flexor contractions; 100-m and 400-m swim times were also recorded within 1-wk of the postsupplement testing	Cr had no effect on metabolite ratios or muscle metabolism as measured by PMRS and had no effect on muscle reoxygenation; Cr also did not improve TTE during the fatiguing plantarflex contractions or swim performance times	Cr supplementation at 2 g daily had no effect on muscle-Cr concentration; muscle oxygen supply, or muscle metabolism during endurance exercise	*17*
Healthy, untrained, nonvegetarian men; age ± SD = 21 ± 1 yr; $n = 20$	Group-matched (based on workload and VO_{2max}), mixed factorial, double-blinded, Pl-controlled design; Cr group	Cr group ingested 20 g/d of Cr for the first 5 d, followed by 37 d of 2 g/d assessments were for muscle-Cr content, body composition,	Five days of Cr loading (20 g/d) increased muscle-Cr and phosphocreatine (PCr), but 2 g/d	Cr loading or long-term Cr supplementation does not influence oxidative	*58*

(Continued)

Table 1 *(Continued)*

Population	*Design*	*Intervention*	*Results*	*Comments*	*References*
	($n = 10$) vs Pl group ($n = 10$), pre-, mid- (after the 7-d loading phase), and posttests were conducted after a 42-d supplementation period	oxidative capacity, and substrate utilization during supramaximal and submaximal cycle ergometry	was not able to maintain high levels of muscle-Cr and -PCr. Cr improved repeated supramaximal cycle sprint performance, but had no effect on oxidative capacity, substrate utilization, or time-trial performance. Cr increased body mass and fat-free mass	capacity or substrate utilization during traditional endurance tasks. However Cr does improve intermittent, repeated, supramaximal sprint cycle endurance performance	
Healthy, welltrained, amateur cyclists,	Double-blinded, Pl-controlled, repeated measures, cross over design;	For Cr loading, 25 g/d of Cr for 5 d; Cr administered during the test was 5 g/h; Pl was	Cycling speed and TTE was not affected by Cr supplementation	Cr loading improved intermittent sprint cycle performance at	*67*

nonvegetarian men; 18–34 yr; $n = 12$	three randomly ordered conditions: (1) Cr-loaded, (2) Cr-loaded + Cr ingestion during the test, and (3) Pl each separated by a 5-wk washout period	glucose + maltodextrine + aspartame; standard cycle ergometer test of 150 min followed by five repeated, intermittent 10-s cycling sprints were performed on four occasions: (1) presupplementaton and (2) postsupplementation after Cr loading, Cr loading + acute ingestion, and Pl. Endurance was assessed with blood lactate at various workloads and TTE at the end of the 150-min ride, whereas sprint performance was assessed by power output declines during the five intermittent sprint tests	during the endurance ride. Cr improved power output (8–9%) and increased plasma Cr during the intermittent sprints. Blood lactate, pH, plasma Cr, urea, and uric acid levels were unaffected by Cr supplementation	the end of a 150-min endurance ride, but performance was slightly diminished by administering Cr during the exercise. Cr did not improve endurance cycling performance	
Healthy men; age ± SD =	Randomized, double-blinded, Pl-controlled,	All supplements were administered four times	After adjusting for differences among	CrBA may enhance endurance	*43*

(Continued)

Table 1 *(Continued)*

Population	*Design*	*Intervention*	*Results*	*Comments*	*References*
25 ± 5 yr; $n = 55$	mixed factorial design; four group: (1) Pl group, (2) Cr group, (3) β-alanine (β-Ala), and (4) Cr + β-alanine (CrBA); testing occurred before and after the 28-d supplementation period	times per day for the first 6 d then twice per day for 22 d. Individual doses were 34 g dextrose (Pl, $n = 13$), 5.25 g Cr + 34 g dextrose (Cr, $n = 12$), 1.6 g β-Ala + 34 g dextrose (β-Ala, $n = 14$), and 5.25 g Cr + 1.6 g β-Ala + 34 g dextrose (CrBA, $n = 16$). Each subject performed incremental cycle ergometry exercise to exhaustion before and after the supplementation period to determine VO_{2peak}; lactate, and VT values for VO_2 workload and $\%VO_{2peak}$ and TTE	the presupplementation means, there were no differences between groups (Pl, Cr, β-Ala, or CrBA) for the postsupplementation means for any of the dependant variables. Dependent-samples *t*-tests showed that CrBA improved LT VO_2 and workload and VT VO_2 workload and $\%VO_{2peak}$ whereas Cr alone improved the VT workload and TTE	performance beyond Cr and β-Ala taken alone	

should be taken). When interpreting the RE values in this review, RE less than 100 indicated that Cr supplementation decreased the endurance performance variable, RE equals to 100 means that Cr supplementation had no RE on the endurance performance variable, and RE more than 100 indicated that Cr supplementation increased the endurance performance variable. For instance, if the RE value was 120, Cr supplementation resulted in a 20% increase in that particular measurement of endurance performance compared with the Pl, control, or pretesting condition. Because it was not possible to calculate the confidence intervals without the raw data, the figures display only the point estimates.

2.1. Study Delimitations

- Of the 35 studies that were reviewed, 24 were randomized, seven were nonrandomized, and four were unable to be classified.
- Thirty-one were Pl-controlled, whereas four studies had no Pl condition.
- Twenty-five of thirty five were a mixed factorial design (two or more groups tested at two or more times), eight were a crossover design (all subjects experienced all conditions with a washout period), and two were repeated measures designs (testing occurred before and after supplementation without a control or Pl group).
- Twenty-six were double-blinded (neither the subjects nor the investigators were aware of supplement groups), seven were nonblinded (everyone knew what was being consumed), and two were single-blinded (the subjects did not know whether they were consuming a Pl or Cr, but the investigators knew).
- Thirty-two of thirty five studies included subjects between the ages of 18 and 37 yr, two studies involved older adults between the ages of 55 and 80 yr, and one study was on young athletes with an average age of 17 yr.
- Twenty-one studies only examined men, nine studies examined both genders, four studies looked at women only, and one was unclassified.
- All studies were published between 1993 and 2006.

3. MEASURES OF ENDURANCE PERFORMANCE

From the 35 studies reviewed, five variables were chosen to represent measures of endurance performance: (a) oxygen consumption, (b) lactate concentrations, (c) time-to-exhaustion (TTE), (d) anaerobic thresholds, and (e) average work and peak power output accomplished.

3.1. Oxygen Consumption

Oxygen (O_2) consumption rate or VO_2 is the body's ability to extract O_2 from the air and use it to synthesize ATP for energy during a process known as oxidative phosphorylation. VO_2 can be measured during maximal and submaximal exercise intensities. Maximal oxygen consumption rate (VO_{2max} or VO_{2peak}) is typically measured during an incremental aerobic exercise (i.e., treadmill, cycle ergometer, or rowing ergometer), wherein the exercise starts at a low intensity and is gradually increased in stages until volitional exhaustion. VO_2 is assessed by analyzing the gases expired from the lungs for O_2 and CO_2 content. VO_{2max} is a direct indicator of *aerobic power* or *aerobic capacity*, and it is a common laboratory measurement to assess cardiorespiratory fitness *(32)*. It is well known that VO_{2max} is strongly and directly correlated with endurance exercise performance *(12)*. In contrast, submaximal VO_2 measurements can provide an index of exercise efficiency. That is, as one adapts to an aerobic exercise training program, less O_2 is consumed at the same absolute level of exercise intensity indicating that less energy was required to perform the same task. In addition, when VO_2 is recorded during the transition from rest to steady-state exercise, there is a 3–4 min delay in O_2 uptake when VO_2 increases rapidly, but not fast enough to account for the energy needed during the first 3 or 4 min of exercise. The amount of energy expended during this transitionary period of exercise is termed the *O_2 deficit*. During the O_2 deficit, the metabolic demands of the exercise are met by the PCr and anaerobic glycolysis energy systems *(33,34)*. Therefore, it has been hypothesized that increasing total muscle-Cr stores as a result of Cr supplementation may delay the reliance on aerobic metabolism, which might increase the O_2 deficit and reduce submaximal VO_2 *(26,35)*. This hypothesis is also consistent with the mechanism of the PCr shuttle. That is, if the ATP:ADP ratio is maintained for longer durations after Cr supplementation, then the accumulation of ADP would not be sufficient to stimulate the ADP-sensitive mitochondrial oxidative phosphorylation.

Of the 35 studies reviewed, nine studies measured the effects of Cr supplementation on VO_{2max} *(35–43)* during exercise durations lasting between 30 s *(36)* and 28 min *(41)*, whereas six studies measured the effects of Cr supplementation on submaximal VO_2 *(14,19,26,35,44,45)* during exercise intensities ranging from approx 20% *(35)* to 90% *(44)* of the estimated maximal intensity. Ironically, only one study *(35)* reported both maximal and submaximal VO_2 values. Figure 1 displays the RE of Cr supplementation on VO_{2max}, where the *y*-axis represents the exercise duration in minutes. Figure 2 shows the RE of Cr supplementation on submaximal VO_2, where the *y*-axis represents exercise intensity.

Based on the RE values plotted in Fig. 1, only two studies *(36,40)* demonstrated a positive effect of Cr supplementation on VO_{2max} (0.2 and 1.5% increase), whereas the other point estimates indicated a

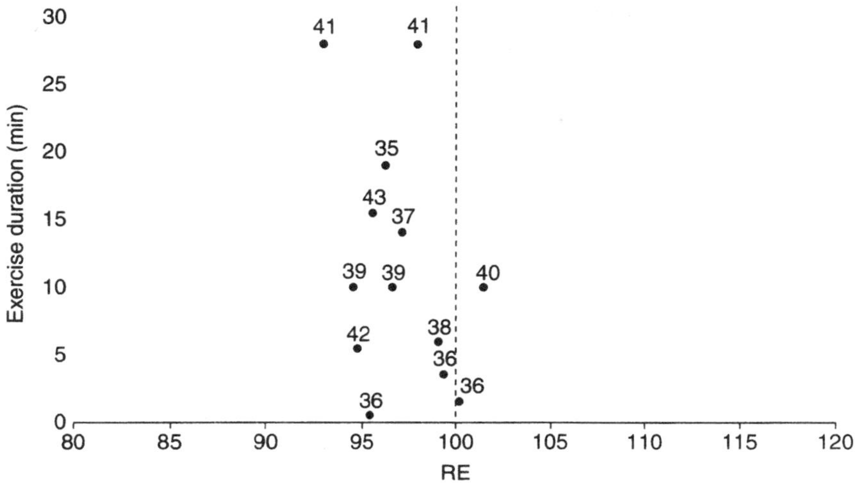

Fig. 1. RE of Cr supplementation on maximal oxygen consumption rate (VO_{2max} or VO_{2peak}) plotted as a function of exercise duration (min). The number nearest the point estimate of RE is the reference number for the study it represents. The vertical dashed line represents the line of equality, where Cr supplementation had no affect on VO_{2max}. RE greater than 100 indicated increases in VO_{2max}, whereas RE less than 100 indicated decreases as a result of Cr supplementation compared with the Pl or control condition.

0.5–7% decrease in VO_{2max} with Cr. The average RE value across all the point estimates was 97.1. Given the margin of error associated with the calculation of the RE point estimates used in this review, these findings indicated that Cr supplementation has little to no effect on VO_{2max}, with a possibility that VO_{2max} might actually be slightly reduced with Cr intake. It has been suggested that the lack of change or slight decreases in aerobic capacity may be because of the increases in body mass that are often observed with Cr supplementation *(1,14,36,41)*; however, this hypothesis has not been confirmed. In addition, there appears to be no relationship between the effects of Cr on VO_{2max} and the duration of exercise used to assess VO_{2max}. These findings are not surprising when considering that factors such as cardiac output, muscle blood flow and oxygenation *(46)*, and O_2 carrying capacity of the blood *(12)* are the primary determinants of VO_{2max}, rather than Cr or CrP stores and related energy systems.

The RE values plotted in Fig. 2 indicated that two studies reported increases (3.5–13.6%) *(44,45)*, two studies reported decreases (8.6–20%) *(35,44)*, and five studies reported equivocal findings (4.1% decrease – 0.5% increase) *(14,19,26,44,45)* in submaximal VO_2 with Cr

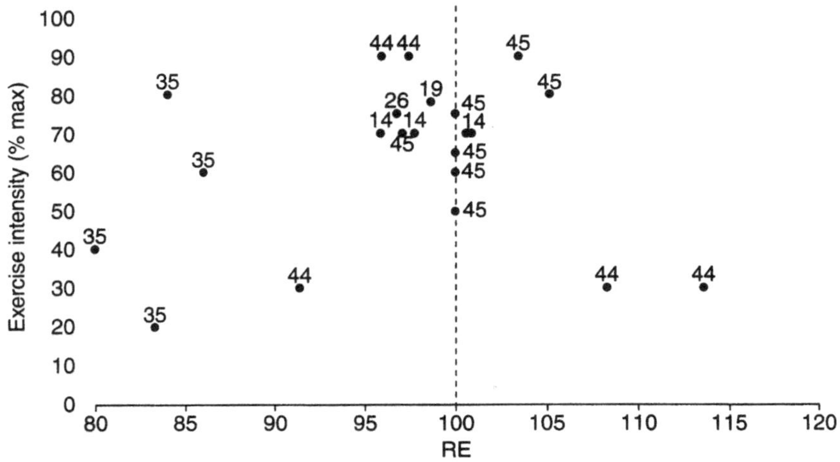

Fig. 2. RE of Cr supplementation on submaximal oxygen consumption rate (VO_2) plotted as a function of exercise intensity (% max). The number nearest the point estimate of RE is the reference number for the study it represents. The vertical dashed line represents the line of equality, where Cr supplementation had no affect on the submaximal VO_2. RE greater than 100 indicated increases in submaximal VO_2, whereas RE less than 100 indicated decreases as a result of Cr supplementation compared with the Pl or control condition.

supplementation. Interestingly, the two highest RE values calculated from Rico-Sanz and Marco *(44)* occurred during active recovery intervals (30% of max) between 10-s high-intensity cycling bouts. The four lowest RE values were calculated from submaximal VO_2 responses during a graded exercise test *(35)*. The unique findings of Rico-Sanz and Marco *(44)* were compared with the similar increases observed in vitro after Cr supplementation *(20–22)*. However, these findings have not been replicated. In contrast, the results of two well-controlled studies *(26,35)* have demonstrated decreases in submaximal VO_2 after Cr supplementation. These findings are consistent with the hypothesis that increases in intramuscular Cr and CrP may delay oxidative phosphorylation by prolonging the rise in ADP concentrations. Nelson et al. *(35)* and Jones et al. *(26)* have both suggested that Cr might improve exercise efficiency, which aligned with the findings of other studies that have demonstrated increases in the anaerobic working capacity (AWC) *(40,47–49)* and the physical working capacity at the neuromuscular fatigue threshold *(50,51)*. Overall, the evidence suggested that Cr supplementation tends to reduce submaximal oxygen uptake at a given workload, which further suggested that Cr might increase submaximal exercise efficiency. However, the effects of Cr supplementation on submaximal VO_2 are somewhat inconclusive, as other studies have demonstrated no appreciable changes *(14,19,26,44,45)*. In addition, there appears to be no visual evidence based on the spread of RE point estimates in Fig. 2 that the Cr-related changes in submaximal VO_2 are related to exercise intensity.

3.2. Lactate Concentrations

In skeletal muscle, lactate molecules are formed in the sarcoplasm by the reduction of pyruvate, a reaction that is catalyzed by the enzyme *lactate dehydrogenase (52)*. Pyruvate is commonly regarded as the product of the first phase of glycolysis, where a six-carbon chain sugar (i.e., glucose, fructose, and so on) is catabolized into two three-carbon chain pyruvate molecules *(52)*. There are several potential "fates" of pyruvate *(11)*; however, the two most common during skeletal muscle metabolism are (a) pyruvate being shuttled into the mitochondria to be further catabolized in the Krebs cycle or (b) pyruvate being converted to lactate *(12)*. During exercise, these two fates of pyruvate are largely determined by the intensity of the exercise. In light to moderate exercise, enough O_2 is available to favor the catabolism of pyruvate in the mitochondria, and lactate concentrations remain low and stable. However, during strenuous exercise, when the energy demands exceed the supply of O_2 as well as the immediate supply of ATP, CrP concentrations decrease in order to rapidly rephosphorylate ADP through the CK reaction (Eq. 1), which also triggers a dramatic increase in the rate of glycolysis to increase ATP availability. Therefore, pyruvate accumulates faster than it can be shuttled into the mitochondria, and it is consequently reduced to lactate. Therefore, lactate concentrations in the muscle or blood are indicators of anaerobic energy production.

CrP may be a strong allosteric inhibitor of the PFK enzyme that regulates glycolysis *(53)*. Therefore, when CrP concentrations decrease during intense exercise, PFK may become less inhibited, which would increase the rate of glycolysis and lactate production *(6)*. However, with Cr supplementation it has been suggested that increases in Cr and CrP concentrations in the muscle may delay the reduction in CrP concentrations during intense exercise, consequently reducing the rate of lactate formation *(6,54)*. This has also been described as a decrease in the reliance on glycolysis to provide ATP, which may improve performance output of the working muscles *(26,54)*; however, the evidence to support this hypothesis is inconclusive *(55)*.

Of the 35 articles reviewed, 15 studies included measures of blood or plasma lactate in response to Cr supplementation either at rest, during exercise, and/or in recovery from exercise. Based on the RE values calculated from these 15 studies (Fig. 3), blood lactate concentrations have been reported to increase *(10,36,45,55–58)*, decrease *(18,26,44,54)*, or remain unchanged *(14,19,37,54,59,60)* as a result of Cr supplementation compared with the Pl or control condition. At rest, the average RE value for blood lactate was 101.8 (range = 83.3–118.8), the average RE value for blood lactate concentrations during exercise was 100.5 (range = 65.6–128.9), and the average RE value for blood lactate recovery from exercise was 105 (range = 86.2–138.4). The average RE values were near 100, initially suggesting that Cr supplementation had no affect on lactate concentrations; however, the ranges were quite large. Therefore, these findings suggested that the effects of Cr supplementation on blood lactate concentrations are not consistent across studies. Prevost et al. *(54)* suggested that the decreases in blood lactate following Cr supplementation may have reduced the reliance on anaerobic glycolysis, which in turn, blunted the drop in pH that allowed an extended TTE during intermittent cycling. Jones et al. *(26)* also demonstrated decreases in blood lactate with Cr supplementation and associated this with an increase in cycling efficiency, which may have been possible because cycling performance is independent of body mass. However, with running and other whole-body exercises like kayaking and rowing, increases in body mass that are typical with Cr loading may increase the energy costs of exercise, which may negate any improvements in efficiency *(26)*. Indeed, the decreases in blood lactate concentrations were observed during cycling exercises *(18,26,44,54)*, whereas the increases in blood lactate were primarily observed during running *(36,45,57)*, rowing *(56)*, or kayaking *(55)*, which supported the hypothesis of Jones et al. *(26)* that increases in body mass may adversely impact efficiency (as assessed by blood lactate concentrations) for modes of exercise that involve the propulsion of the body. This hypothesis was supported by Branch *(5)* who indicated that "Bicycle ergometry was the only mode of aerobic exercise for which a significant ES (effect size) was observed" (p. 216). Based on the evidence in the present review, measures of blood lactate may not be stable enough to consistently demonstrate any beneficial effects of Cr supplementation; however, with modes of exercise that are independent of body mass (i.e., cycling), Cr may improve efficiency by reducing the reliance on anaerobic glycolysis.

3.3. Time-to-Exhaustion

Perhaps the most simple and obvious measure of endurance performance is the TTE. TTE is measured as the time to volitional exhaustion

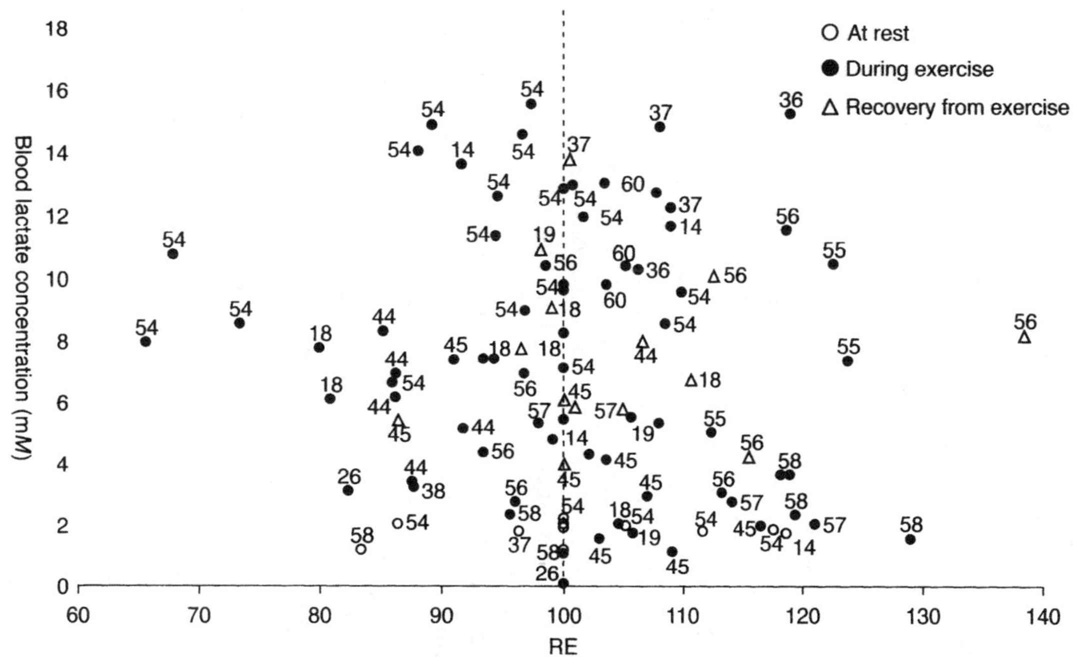

Fig. 3. RE of Cr supplementation on blood lactate concentrations plotted as a function of blood lactate concentration (m*M*). The number nearest the point estimate of RE is the reference number for the study it represents. The vertical dashed line represents the line of equality, where Cr supplementation had no affect on lactate concentrations. RE greater than 100 indicated increases in lactate concentrations, whereas RE less than 100 indicated decreases as a result of Cr supplementation, compared with the Pl or control condition.

during either a constant work rate or incremental exercise test. The longer the TTE, the greater the resistance to fatigue, which allows for improvements in endurance performance. It has been suggested *(54)* that Cr supplementation may increase the TTE by increasing intramuscular Cr and CrP that prolong energy synthesis through the CK reaction (Eq. 1). However, because the CrP energy system contributes most during anaerobic high-intensity exercise, it is possible that Cr supplementation may have a greater effect on TTE for short-durations, rather than lower intensity submaximal workloads that allow longer TTE.

Twenty *(16–19,35,36,41,43,44,54,56,57,60–67)* of the thirty five studies reviewed included some measurement of TTE. The RE point estimates in Fig. 4 showed that as the duration of the TTE decreased, the effects of Cr supplementation on TTE increased. The average RE value for the 0.4–3.6 min durations was 129.4. For the medium durations (4.6–11.8 min), the average RE value was 106.4, whereas for the longer durations (13.7–36.3 min) the average RE was 97. The trend indicated that Cr supplementation was most effective for increasing TTE during short-duration high-intensity exercise lasting up to 3 or 4 min *(17,36,54,56,60,62,65)*, less effective for durations lasting 5–12 min *(16,17,63,65–67)*, but ineffective for exhaustive exercise durations greater than 12 min *(18,19,35,36,41,43,44,57,61)*. This is perhaps not surprising, because the CrP energy system seems to contribute most during high-intensity, anaerobic exercise. In fact, a recent meta-analysis *(5)* used classifications of exercise durations ≤30 s, >30 to ≤150 s, and >150 s, as representations of the CrP energy system, anaerobic glycolysis, and oxidative phosphorylation, respectively. Branch *(5)* concluded that "...the ergogenic potential of Cr supplementation diminishes with increasing duration of activity" (p. 216). Indeed the evidence in the present review supported the findings of Branch *(5)* and indicated that Cr supplementation may be most beneficial for exhaustive exercise efforts lasting up to 3 or 4 min, with some positive effects for exercise durations between 5 and 12 min. However, as the duration of the endurance event increased above 12 min, Cr supplementation had no relative affect on the TTE.

3.4. Anaerobic Thresholds

The effects of Cr supplementation have been examined using several types of anaerobic thresholds, including the ventilatory threshold (VT) *(35,38,39,43)*, the physical working capacity at fatigue threshold (PWC_{FT}) *(50,51)*, the lactate threshold (LT) *(43,56)*, critical power (CP) *(40,65)*, and the AWC *(40,47–49,65)*. The VT is typically determined as the point of nonlinearity or *breaking point* in the relationship

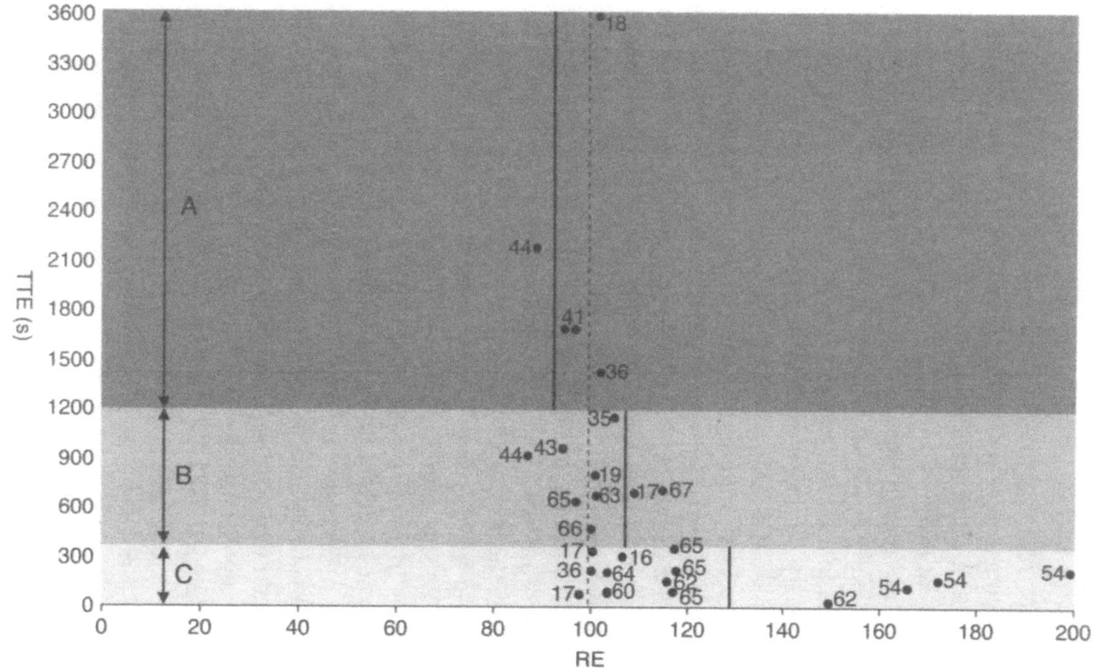

Fig. 4. RE of Cr supplementation on TTE plotted as a function of TTE (s). (A) 820–2180 s durations, (B) 277–710 s durations, and (C) 25–218 s durations. The dark vertical lines represent the average RE values for the A (97), B (106.4), and C (129.4) durations. The number nearest the point estimate of RE is the reference number for the study it represents. The vertical dashed line represents the line of equality, where Cr supplementation had no affect on TTE. RE greater than 100 indicated increases in TTE, whereas RE less than 100 indicated decreases as a result of Cr supplementation compared with the Pl or control condition.

between ventilation and VO_2 *(68)*. The VT represents the point at which ventilation begins to increase disproportionately with VO_2 during an incremental exercise test, which is the result of an increase in anaerobic metabolism *(12)*. If Cr supplementation can delay the contributions of anaerobic metabolism by maintaining the ATP:ADP and/or CrP:Cr ratios, then the VT may increase *(35)*.

The PWC_{FT} represents the power output during cycle ergometry that can be maintained without an increase in muscle activation (i.e., electromyographic amplitude) *(69–74)*. In theory, progressive increases in workload result in the recruitment of higher-threshold fast-twitch motor units that eventually leads to neuromuscular fatigue. However, steady-state workloads that do not increase muscle activation could be maintained continuously without neuromuscular fatigue *(70)*. It is likely that the motor units that are recruited at workloads just above the PWC_{FT} probably have both glycolytic and oxidative properties (i.e., type IIa fibers). Therefore, increases in muscle-Cr and -CrP that occur with Cr supplementation may be able to prolong the fatigue of type IIa fibers by delaying the accumulation of Cr, ADP, and P_i that are potent activators of anaerobic glycolysis. Consequently, Cr supplementation may improve the PWC_{FT}. In addition, because the PWC_{FT} is determined during cycle ergometry *(70)*, increases in body mass may not have any adverse influences on exercise efficiency, which may allow the PWC_{FT} to act as a valid, reliable, and sensitive indicator of the ergogenic benefits of Cr supplementation.

As discussed earlier in Section 3.2., lactate concentrations stay low and stable during low- to moderate-intensity exercise (up to ~50% of the VO_{2max}). As the workload is increased, there is a disproportionate increase in lactate concentrations that represents a greater contribution of anaerobic metabolism in a similar fashion to ventilation and the VT. However, when the point of lactate accumulation is determined, the resultant threshold is termed the LT *(75)*. The VT and LT are similar in many respects, because ventilation will increase in order to "blow off" the CO_2 that forms as a result of lactate buffered by sodium bicarbonate *(12)*. Thus, the VT can be considered a less direct measure of anaerobic metabolism than the LT, as the LT involves calculations directly from lactate concentrations. The conclusions drawn from the evidence in Sections 3.2. and 3.3. indicated that the effects of Cr supplementation on VO_2 (derived from ventilation and expired gases) and lactate concentrations were somewhat inconclusive and quite variable; therefore, it is unclear how Cr supplementation may affect the VT and LT, because both thresholds are at least partially based on either VO_2 or lactate measures.

The relationship between work rate and TTE has traditionally been characterized as a curvilinear, hyperbolic function *(76)*. Previous studies have used various mathematical models to predict and describe the parameters of this relationship *(77,78)*. Perhaps the simplest model is the linear equation *(65)*:

$$W_{\text{lim}} = \dot{W}_{\text{CP}}\,(t) + W' \qquad (6)$$

where W_{lim} is the total amount of work performed during an exercise bout, $\dot{W}_{\text{CP}}$ is the slope, t is the TTE, and W' is the y-intercept. Consequently, the slope ($\dot{W}_{\text{CP}}$) has been termed CP and has been described as the workload that can be sustained indefinitely without fatigue *(76)* as well as an inherent representation of the aerobic, oxidative energy systems *(79)*. In contrast, the y-intercept of this relationship (W') has been called the AWC, which is said to reflect the anaerobic energy reserve of the muscle *(76,80)*. Therefore, as the CP is closely associated with aerobic energy synthesis, whereas the AWC is reflective of anaerobic metabolism, it is not surprising that Cr supplementation may improve the AWC, but not the CP *(40,65)*. Furthermore, t is often determined for workloads that elicit volitional exhaustion within 6 *(40)*, 10 *(47–49)*, or 12 *(65)* min. The conclusion from Section 3.3. was that Cr supplementation may slightly improve the TTE when the resultant durations are 5–12 min, but has profound effects when the intensity is high enough to cause exhaustion within 4 min. Thus, Cr may not affect the slope of the linear relationship between work rate and TTE (i.e., CP), but it may shift the entire relationship upward, which would increase the y-intercept (i.e., AWC).

Twelve *(35,38–40,43,47–51,56,65)* of the thirty five studies included at least one of the anaerobic thresholds (VT, PWC_{FT}, LT, CP, or AWC). Based on the evidence in Fig. 5, the PWC_{FT} *(50,51)* and AWC *(40,47–49,65)* thresholds benefited most from Cr supplementation, whereas the VT *(35,38,39,43)*, LT *(43,56)*, and CP *(40,65)* were relatively unaffected. The VT and LT values were calculated based on relationships between ventilation vs VO_2 and lactate vs workload, respectively. Based on the evidence presented in Sections 3.1. and 3.2., it is not surprising that these thresholds were not substantially affected by Cr supplementation, simply because VO_2 and lactate concentrations were also not consistently affected by Cr supplementation. One could argue that Cr supplementation may have improved the VT in the study by Nelson et al. *(35)*; however, the other four estimates of VT were unaltered. It is also possible that Cr supplementation may have slightly improved the LT, because both RE values in Fig. 5 (105.4 and 108) were determined with a cycle ergometer *(43,56)*. This finding was

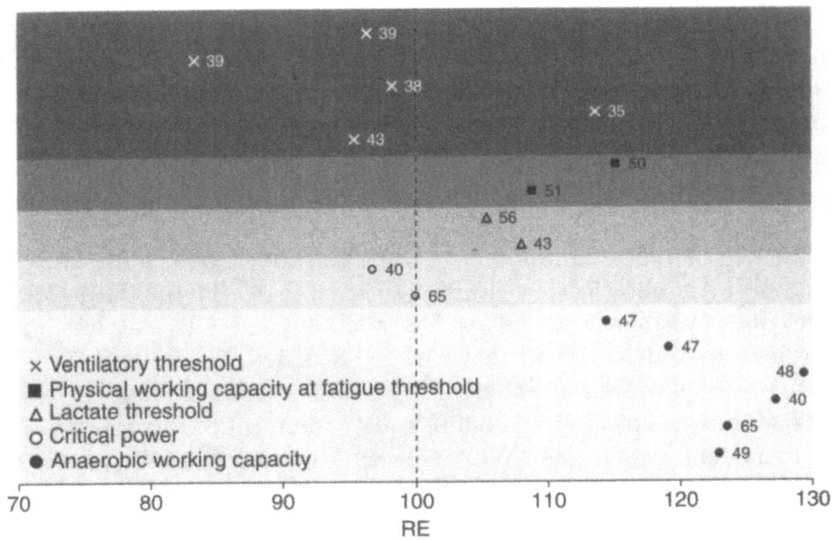

Fig. 5. RE of Cr supplementation on various anaerobic thresholds. The number nearest the point estimate of RE is the reference number for the study it represents. The order of placement on the *y*-axis was arbitrary, but the RE values were grouped according to the type of threshold. The vertical dashed line represents the line of equality, where Cr supplementation had no affect on the threshold. RE greater than 100 indicated increases in the threshold value, whereas RE less than 100 indicated decreases as a result of Cr supplementation compared with the Pl or control condition.

consistent with the previous hypothesis that cycle ergometry exercise may be able to exploit the benefits of Cr supplementation, because the workload is independent of body mass. Because CP predominantly represents the contributions of the aerobic energy systems to exercise, it is not surprising that this slope value does not appreciably change with Cr supplementation. However, the Cr-induced upward shift of the relationship between work rate and TTE does seem to improve the *y*-intercept (i.e., AWC). Overall, from the evidence in Fig. 5, it appears that the ergogenic benefits of Cr supplementation are most noticeable in the PWC_{FT}, AWC, and possibly the LT when calculated from lactate concentrations acquired during cycle ergometry. Interestingly, each of these "thresholds" (PWC_{FT}, AWC, and LT) were determined during cycle ergometry, which is a nonweight bearing mode of exercise that may not be confounded by the potential increases in body mass that are typically observed with Cr supplementation.

3.5. Average Work and Peak Power Output Accomplished

Additional simplistic measurements of endurance performance include the average work and peak power output accomplished during an exercise bout, which are similar to the TTE measurement. The peak power output can be quantified during a maximal test to exhaustion *(38,56)* or during a fixed interval time trial *(42,55,58)*. Most often, the average work accomplished is measured during a time trial *(42,55,58)*. In these studies, the average work performed or maximal power output achieved is quantified after the participant is asked to "work as hard as possible." It should be noted that a TTE test by definition is a maximal test that requires the participant to reach failure, whereas, a time trial implies a pacing strategy that may not necessarily elicit complete exhaustion. That said, the peak power outputs reported by Canete et al. *(38)* and Chwalbinska-Moneta *(56)* were the results of incremental exercise tests to exhaustion. The average work and peak power outputs reported by McNaughton et al. *(55)*, Reardon et al. *(42)*, and van Loon et al. *(58)* were recorded during time trials. In one study *(58)*, the time trials were reported to have "…a much smaller variation…" (p. 155) than the open-ended trials to exhaustion.

The relationship between *work* and *power* is defined by the following equations *(11)*:

$$\text{Power} = \frac{\text{Work}}{\text{Time}} \tag{7}$$

$$\text{Work} = \text{Force} \times \text{Distance} \tag{8}$$

In Eq. 8, *Force* is the strength or force applied to the ergometer, whereas *Distance* is the distance traveled as a result of the force applied. In Eq. 7, *Time* is the duration of time over which the *Work* was quantified. The units used to measure work and power in the studies reviewed were kilojoules (kJ) and watts (W), respectively. Based on Eqs. 7 and 8, one would assume that if work and power were quantified during a time trial, and the time remained the same, that both work and power would respond the same to a Cr supplementation intervention. Indeed this would be true; however, peak power output is recorded as the highest power output achieved somewhere during the time trial, whereas the average work is typically calculated across the entire time trial. Therefore, when asked to perform as hard and fast as possible, peak power output is usually near the beginning of the time trial and is dependent on the force applied, the cadence, and a discrete time interval, whereas average work is a representation of the entire work bout.

The average work or peak power output accomplished during a time trial is perhaps the most practically applied measurement of performance for athletic competition, as most sport events include a fixed distance. That is, if an increase in average work or peak power output can be demonstrated as a result of Cr supplementation, this could be directly translated to sports performance. On the other hand, few (if any) sports are determined by open-ended competitions where the last person to fail wins, such as the case with tests for TTE.

Five of the thirty five studies reported some measurement of average work *(42,55,58)* or peak power output *(38,42,55,56,58)*. Figure 6 shows the point estimates for RE for average work (filled circles) and peak power output (open circles). The studies by Reardon et al. *(42)* and van Loon et al. *(58)* showed slight improvements in both average work and peak power output during the 15- and 20-min cycling time trials, respectively. Interestingly, both work and power measurements responded similarly *(42,58)*. McNaughton et al. *(55)* reported markedly better average work than peak power during kayaking on an ergometer as a result of Cr supplementation. In fact, much like the results of Section 3.3., there was an inverse relationship between the duration of the time trial and the RE of Cr supplementation. For the short-duration (i.e., 90-s) time trials, Cr resulted in a substantial improvement in average work, whereas the RE of Cr supplementation decreased as the duration of the time trial increased (i.e., 150 and 300 s). However, at the same time, McNaughton et al. *(55)* showed only marginal improvements in peak power. Peak power was greater after Cr supplementation in the Canete et al. *(38)* study; however, this effect was not reported as statistically significant. There was no appreciable effect of Cr on peak power in the study by Chwalbinska-Moneta *(56)*. Overall, it may be important to recognize that all RE estimates were greater than 100 in Fig. 6, which suggested that Cr supplementation can improve average work and peak power output. Improvements in peak power output were only marginal, whereas average work may have been more affected by Cr supplementation.

4. CONCLUSION

From an applied perspective, this review sought to determine whether Cr supplementation improves endurance performance. Based on the collective evidence presented, Cr may improve endurance, but the magnitude of improvement seems to be dependent on two key issues: (a) the duration of the endurance event, which in most cases is dictated by the intensity of exercise and (b) the mode of exercise. From

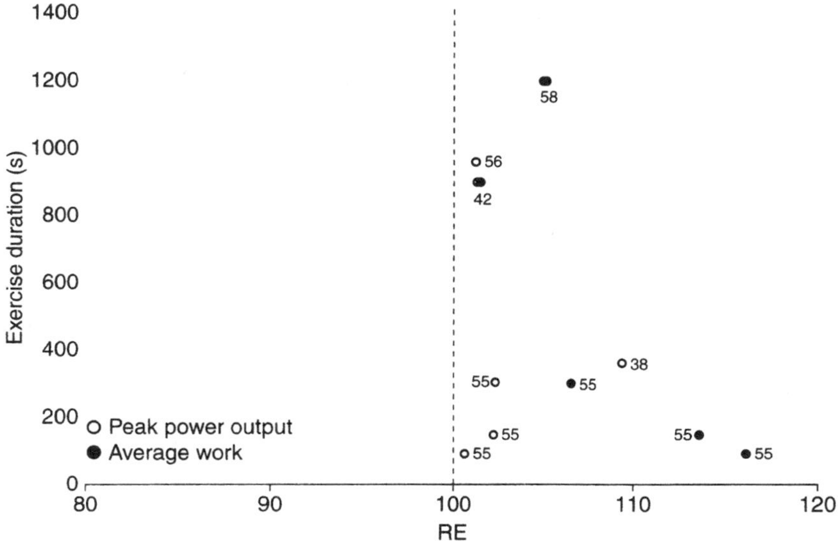

Fig. 6. RE of Cr supplementation on average work and peak power output plotted as a function of exercise duration (s). The number nearest the point estimate of RE is the reference number for the study it represents. The vertical dashed line represents the line of equality, where Cr supplementation had no affect on average work or peak power output. RE greater than 100 indicated increases in performance, whereas RE less than 100 indicated decreases as a result of Cr supplementation compared with the Pl or control condition.

the measures of TTE and average work accomplished (Sections 3.3. and 3.5.. respectively). the RE of Cr supplementation are greatest for short-duration, high-intensity endurance events that last up to approx 3–4 min. The relative ergogenic benefits of Cr seem to diminish as the duration increases (Figs. 4 and 6), such that endurance events much greater than 12 min seem to exhibit no beneficial effects from Cr supplementation. These findings are not necessarily surprising given the anaerobic nature of the CrP energy system. Indeed, Cr supplementation may increase muscle-Cr and -CrP stores, which likely maintains the ratios of CrP:Cr and ATP:ADP that prolong the contributions of anaerobic glycolysis, decreases in pH, lactate accumulation, and subsequent fatigue. In addition, the mode of exercise may also dictate the potential ergogenic benefits of Cr. For example, the performance of exercises such as running, kayaking, and rowing that involve the propulsion of the body may be adversely affected by the gains in body mass that are typically observed with Cr supplementation. Increases in body mass may wash out the beneficial effects of Cr by increasing the energy cost of exercise during weight-bearing activities. However, exercises that are nonweight bearing, such as cycling or swimming, may allow for Cr-induced increases in exercise efficiency.

From a more basic perspective, the variables used to assess the RE of Cr supplementation on endurance performance seem to be important. For example, Cr supplementation had no appreciable affects (or even elicited decreases) on VO_{2max} or CP (Figs. 1 and 5). VO_{2max} is a common and highly-regarded assessment of aerobic capacity and endurance performance, and CP is a sensitive index of aerobic oxidative energy contributions to exercise; therefore, these variables would not be recommended for assessing the impact of Cr supplementation on endurance performance. Variables such as LT and peak power output were only marginally improved by Cr supplementation (Figs. 5 and 6), and as such, should be interpreted with caution when determining the impact of Cr supplementation on endurance performance. Lactate concentrations and VT exhibited a wide range of responses after Cr supplementation (Figs. 3 and 5), which might be interpreted as inconclusive. Again, however, the effects of Cr on lactate concentrations may be dependent on the mode of exercise, with decreases in lactate being most noticeable during cycle ergometry. The variables most profoundly influenced by Cr supplementation included TTE (Fig. 4), the PWC_{FT}, AWC (Fig. 5), average work (Fig. 6), and to some extent, submaximal VO_2 (Fig. 2). The RE of Cr supplementation on TTE was inversely proportional to the duration (and subsequent intensity) of the endurance task. Therefore, when choosing a Cr-sensitive TTE index, an intensity sufficient for eliciting exhaustion within 3 or 4 min is recommended. The PWC_{FT}, average work, and submaximal VO_2 variables all seem to reflect improvements in exercise efficiency and delayed fatigue responses that are expected with Cr supplementation, whereas the AWC reflects the anaerobic energy reserve. However, it should be noted that cycle ergometry seems to be the most effective mode of exercise for assessing the RE of Cr. Therefore, TTE, PWC_{FT}, AWC, average work, or submaximal VO_2 (with careful considerations of exercise duration and mode) are recommended as sensitive indexes to exploit the ergogenic benefits of Cr supplementation for endurance performance.

REFERENCES

1. Demant TW, Rhodes EC. Effects of creatine supplementation on exercise performance. Sports Med 1999; 28:49–60.
2. Bemben MG, Lamont HS. Creatine supplementation and exercise performance: recent findings. Sports Med 2005; 35:107–125.

3. Juhn MS, Tarnopolsky M. Oral creatine supplementation and athletic performance: a critical review. Clin J Sport Med 1998; 8:286–297.
4. Kraemer WJ, Volek JS. Creatine supplementation. Its role in human performance. Clin Sports Med 1999; 18:651–666, ix.
5. Branch JD. Effect of creatine supplementation on body composition and performance: a meta-analysis. Int J Sport Nutr Exerc Metab 2003; 13:198–226.
6. Volek JS, Kraemer WJ. Creatine supplementation: Its effect on human muscular performance and body composition. J Strength Cond Res 1996; 10:200–210.
7. Needham DM. Machina carnis: the biochemistry of muscular contraction in its historical development, Cambridge University Press, Cambridge, 1971.
8. Harris RC, Soderlund K, Hultman E. Elevation of creatine in resting and exercised muscle of normal subjects by creatine supplementation. Clin Sci (Lond) 1992; 83:367–374.
9. Tesch PA, Thorsson A, Fujitsuka N. Creatine phosphate in fiber types of skeletal muscle before and after exhaustive exercise. J Appl Physiol 1989; 66:1756–1759.
10. Greenhaff PL, Bodin K, Soderlund K, Hultman E. Effect of oral creatine supplementation on skeletal muscle phosphocreatine resynthesis. Am J Physiol 1994; 266:E725–E730.
11. DeVries HA, Housh TJ. Physiology of exercise for physical education, athletics, and exercise science. 5th edition, WCB Brown & Benchmark, Madison, Wis, 1994.
12. McArdle WD, Katch FI, Katch, V. L. Exercise physiology: energy, nutrition, and human performance. 6th edition, Lippincott Williams & Wilkins, Baltimore, 2006.
13. Bessman SP, Geiger PJ. Transport of energy in muscle: the phosphorylcreatine shuttle. Science 1981; 211:448–452.
14. Engelhardt M, Neumann G, Berbalk A, Reuter I. Creatine supplementation in endurance sports. Med Sci Sports Exerc 1998; 30:1123–1129.
15. Kemp GJ, Taylor DJ. Styles P, Radda GK. The production, buffering and efflux of protons in human skeletal muscle during exercise and recovery. NMR Biomed 1993; 6:73–83.
16. Smith SA, Montain SJ, Matott RP, Zientara GP, Jolesz FA, Fielding RA. Effects of creatine supplementation on the energy cost of muscle contraction: a 31P-MRS study. J Appl Physiol 1999; 87:116–123.
17. Thompson CH, Kemp GJ, Sanderson AL, et al. Effect of creatine on aerobic and anaerobic metabolism in skeletal muscle in swimmers. Br J Sports Med 1996; 30:222–225.
18. Bellinger BM, Bold A, Wilson GR, Noakes TD, Myburgh KH. Oral creatine supplementation decreases plasma markers of adenine nucleotide degradation during a 1-h cycle test. Acta Physiol Scand 2000; 170:217–224.
19. McConell GK, Shinewell J, Stephens TJ, Stathis CG, Canny BJ, Snow RJ. Creatine supplementation reduces muscle inosine monophosphate during endurance exercise in humans. Med Sci Sports Exerc 2005; 37:2054–2061.
20. Seraydarian MW, Artaza L, Abbott BC. Creatine and the control of energy metabolism in cardiac and skeletal muscle cells in culture. J Mol Cell Cardiol 1974; 6:405–413.
21. Veksler VI, Kuznetsov AV, Anflous K, et al. Muscle creatine kinase-deficient mice. II. Cardiac and skeletal muscles exhibit tissue-specific adaptation of the mitochondrial function. J Biol Chem 1995; 270:19,921–19,929.
22. Saks VA, Rosenshtraukh LV, Smirnov VN, Chazov EI. Role of creatine phosphokinase in cellular function and metabolism. Can J Physiol Pharmacol 1978; 56:691–706.

23. Schneider C, Stull GA, Apple FS. Kinetic characterization of human heart and skeletal muscle CK isoenzymes. Enzyme 1988; 39:220–226.
24. Huso ME, Hampl JS, Johnston CS, Swan PD. Creatine supplementation influences substrate utilization at rest. J Appl Physiol 2002; 93:2018–2022.
25. Wu F, Jeneson JA, Beard DA. Oxidative ATP Synthesis in Skeletal Muscle is Controlled by Substrate Feedback. Am J Physiol Cell Physiol 2006.
26. Jones AM, Carter H, Pringle JS, Campbell IT. Effect of creatine supplementation on oxygen uptake kinetics during submaximal cycle exercise. J Appl Physiol 2002; 92:2571–2577.
27. Kuznetsov AV, Tiivel T, Sikk P, et al. Striking differences between the kinetics of regulation of respiration by ADP in slow-twitch and fast-twitch muscles in vivo. Eur J Biochem 1996; 241:909–915.
28. Tonkonogi M, Harris B, Sahlin K. Mitochondrial oxidative function in human saponin-skinned muscle fibres: effects of prolonged exercise. J Physiol 1998; 510(Pt 1):279–286.
29. Pulido SM, Passaquin AC, Leijendekker WJ, Challet C, Wallimann T, Ruegg UT. Creatine supplementation improves intracellular Ca2+ handling and survival in mdx skeletal muscle cells. FEBS Lett 1998; 439:357–362.
30. van Leemputte M, Vandenberghe K, Hespel P. Shortening of muscle relaxation time after creatine loading. J Appl Physiol 1999; 86:840–844.
31. Shrier I. Does stretching improve performance? A systematic and critical review of the literature. Clin J Sport Med 2004; 14:267–273.
32. American College of Sports Medicine., Kaminsky LA, Bonzheim KA, American College of Sports Medicine. ACSM's resource manual for Guidelines for exercise testing and prescription. 5th edition, Lippincott Williams & Wilkins, Baltimore, MD, 2006.
33. Karlsson J. Lactate and phosphagen concentrations in working muscle of man with special reference to oxygen deficit at the onset of work. Acta Physiol Scand Suppl 1971; 358:1–72.
34. Sahlin K, Ren JM, Broberg S. Oxygen deficit at the onset of submaximal exercise is not due to a delayed oxygen transport. Acta Physiol Scand 1988; 134:175–180.
35. Nelson AG, Day R, Glickman-Weiss EL, Hegsted M, Kokkonen J, Sampson B. Creatine supplementation alters the response to a graded cycle ergometer test. Eur J Appl Physiol 2000; 83:89–94.
36. Balsom PD, Harridge SD, Soderlund K, Sjodin B, Ekblom B. Creatine supplementation per se does not enhance endurance exercise performance. Acta Physiol Scand 1993; 149:521–523.
37. Barnett C, Hinds M, Jenkins DG. Effects of oral creatine supplementation on multiple sprint cycle performance. Aust J Sci Med Sport 1996; 28:35–39.
38. Canete S, San Juan AF, Perez M, et al. Does creatine supplementation improve functional capacity in elderly women? J Strength Cond Res 2006; 20:22–28.
39. Eijnde BO, Van Leemputte M, Goris M, et al. Effects of creatine supplementation and exercise training on fitness in men 55-75 yr old. J Appl Physiol 2003; 95:818–828.
40. Miura A, Kino F, Kajitani S, Sato H, Fukuba Y. The effect of oral creatine supplementation on the curvature constant parameter of the power-duration curve for cycle ergometry in humans. Jpn J Physiol 1999; 49:169–174.

41. Murphy AJ, Watsford ML, Coutts AJ, Richards DA. Effects of creatine supplementation on aerobic power and cardiovascular structure and function. J Sci Med Sport 2005; 8:305–313.
42. Reardon TF, Ruell PA, Fiatarone Singh MA, Thompson CH, Rooney, K. B. Creatine supplementation does not enhance submaximal aerobic training adaptations in healthy young men and women. Eur J Appl Physiol 2006; 98:234–241.
43. Zoeller RF, Stout JR, O'Kroy JA, Torok DJ, Mielke M. Effects of 28 days of beta-alanine and creatine monohydrate supplementation on aerobic power, ventilatory and lactate thresholds, and time to exhaustion. Amino Acids 2006.
44. Rico-Sanz J, Mendez Marco MT. Creatine enhances oxygen uptake and performance during alternating intensity exercise. Med Sci Sports Exerc 2000; 32:379–385.
45. Stroud MA, Holliman D, Bell D, Green AL, Macdonald IA, Greenhaff PL. Effect of oral creatine supplementation on respiratory gas exchange and blood lactate accumulation during steady-state incremental treadmill exercise and recovery in man. Clin Sci (Lond) 1994; 87:707–710.
46. Gonzalez-Alonso J, Calbet JA. Reductions in systemic and skeletal muscle blood flow and oxygen delivery limit maximal aerobic capacity in humans. Circulation 2003; 107:824–830.
47. Eckerson JM, Stout JR, Moore GA, et al. Effect of creatine phosphate supplementation on anaerobic working capacity and body weight after two and six days of loading in men and women. J Strength Cond Res 2005; 19:756–763.
48. Eckerson JM, Stout JR, Moore GA, Stone NJ, Nishimura K, Tamura K. Effect of two and five days of creatine loading on anaerobic working capacity in women. J Strength Cond Res 2004; 18:168–173.
49. Stout JR, Eckerson JM, Housh TJ, Ebersole KT. The effects of creatine supplementation on anaerobic working capacity. J Strength Cond Res 1999; 13:135–138.
50. Stout J, Eckerson J, Ebersole K, et al. Effect of creatine loading on neuromuscular fatigue threshold. J Appl Physiol 2000; 88:109–112.
51. Stout JR, Cramer JT, Mielke M, O'Kroy J, Torok D, Zoeller RF. Effects of twenty-eight days of beta-alanine and creatine monohydrate supplementation on the physical working capacity at neuromuscular fatigue threshold. J Strength Cond Res 2006; 20, in press.
52. Garrett R, Grisham CM. Biochemistry. 2nd edition, Saunders College Pub., Fort Worth, 1999.
53. Storey KB, Hochachka PW. Activation of muscle glycolysis: a role for creatine phosphate in phosphofructokinase regulation. FEBS Lett 1974; 46:337–339.
54. Prevost MC, Nelson AG, Morris GS. Creatine supplementation enhances intermittent work performance. Res Q Exerc Sport 1997; 68:233–240.
55. McNaughton LR, Dalton B, Tarr J. The effects of creatine supplementation on high-intensity exercise performance in elite performers. Eur J Appl Physiol Occup Physiol 1998; 78:236–240.
56. Chwalbinska-Moneta J. Effect of creatine supplementation on aerobic performance and anaerobic capacity in elite rowers in the course of endurance training. Int J Sport Nutr Exerc Metab 2003; 13:173–183.
57. Izquierdo M, Ibanez J, Gonzalez-Badillo JJ, Gorostiaga EM. Effects of creatine supplementation on muscle power, endurance, and sprint performance. Med Sci Sports Exerc 2002; 34:332–343.

58. van Loon LJ, Oosterlaar AM, Hartgens F, Hesselink MK, Snow RJ, Wagenmakers AJ. Effects of creatine loading and prolonged creatine supplementation on body composition, fuel selection, sprint and endurance performance in humans. Clin Sci (Lond) 2003; 104:153–162.
59. Birch R, Noble D, Greenhaff PL. The influence of dietary creatine supplementation on performance during repeated bouts of maximal isokinetic cycling in man. Eur J Appl Physiol Occup Physiol 1994; 69:268–276.
60. Earnest CP, Almada AL, Mitchell TL. Effects of creatine monohydrate ingestion on intermediate duration anaerobic treadmill running to exhaustion. J Strength Cond Res 1997; 11:234–238.
61. Jakobi JM, Rice CL, Curtin SV, Marsh GD. Contractile properties, fatigue and recovery are not influenced by short-term creatine supplementation in human muscle. Exp Physiol 2000; 85:451–460.
62. Maganaris CN, Maughan RJ. Creatine supplementation enhances maximum voluntary isometric force and endurance capacity in resistance trained men. Acta Physiol Scand 1998; 163:279–287.
63. Ostojic SM. Creatine supplementation in young soccer players. Int J Sport Nutr Exerc Metab 2004; 14:95–103.
64. Rossiter HB, Cannell ER, Jakeman PM. The effect of oral creatine supplementation on the 1000-m performance of competitive rowers. J Sports Sci 1996; 14:175–179.
65. Smith JC, Stephens DP, Hall EL, Jackson AW, Earnest CP. Effect of oral creatine ingestion on parameters of the work rate-time relationship and time to exhaustion in high-intensity cycling. Eur J Appl Physiol Occup Physiol 1998; 77:360–365.
66. Syrotuik DG, Game AB, Gillies EM, Bell GJ. Effects of creatine monohydrate supplementation during combined strength and high intensity rowing training on performance. Can J Appl Physiol 2001; 26:527–542.
67. Vandebuerie F, Vanden Eynde B, Vandenberghe K, Hespel P. Effect of creatine loading on endurance capacity and sprint power in cyclists. Int J Sports Med 1998; 19:490–495.
68. Orr GW, Green HJ, Hughson RL, Bennett GW. A computer linear regression model to determine ventilatory anaerobic threshold. J Appl Physiol 1982; 52:1349–1352.
69. deVries HA, Housh TJ, Johnson GO, et al. Factors affecting the estimation of physical working capacity at the fatigue threshold. Ergonomics 1990; 33:25–33.
70. deVries HA, Tichy MW, Housh TJ, Smyth KD, Tichy AM, Housh DJ. A method for estimating physical working capacity at the fatigue threshold (PWCFT). Ergonomics 1987; 30:1195–1204.
71. Helal JN, Guezennec CY, Goubel F. The aerobic-anaerobic transition: re-examination of the threshold concept including an electromyographic approach. Eur J Appl Physiol Occup Physiol 1987; 56:643–649.
72. Housh TJ, deVries HA, Johnson GO, Evans SA, McDowell S. The effect of ammonium chloride and sodium bicarbonate ingestion on the physical working capacity at the fatigue threshold. Eur J Appl Physiol Occup Physiol 1991; 62:189–192.
73. Housh TJ, deVries HA, Johnson GO, et al. The effect of glycogen depletion and supercompensation on the physical working capacity at the fatigue threshold. Eur J Appl Physiol Occup Physiol 1990; 60:391–394.
74. Viitasalo JT, Luhtanen P, Rahkila P, Rusko H. Electromyographic activity related to aerobic and anaerobic threshold in ergometer bicycling. Acta Physiol Scand 1985; 124:287–293.

75. Weltman A, Snead D, Stein P, et al. Reliability and validity of a continuous incremental treadmill protocol for the determination of lactate threshold, fixed blood lactate concentrations, and VO2max. Int J Sports Med 1990; 11:26–32.
76. Monod H, Scherrer J. The work capacity of a synergic muscular group. Ergonomics 1965; 8:329–338.
77. Gaesser GA, Carnevale TJ, Garfinkel A, Walter DO, Womack CJ. Estimation of critical power with nonlinear and linear models. Med Sci Sports Exerc 1995; 27:1430–1438.
78. Housh TJ, Cramer JT, Bull AJ, Johnson GO, Housh DJ. The effect of mathematical modeling on critical velocity. Eur J Appl Physiol 2001; 84:469–475.
79. Gaesser GA, Wilson LA. Effects of continuous and interval training on the parameters of the power-endurance time relationship for high-intensity exercise. Int J Sports Med 1988; 9:417–421.
80. Nebelsick-Gullett LJ, Housh TJ, Johnson GO, Bauge SM. A comparison between methods of measuring anaerobic work capacity. Ergonomics 1988; 31:1413–1419.

4 Creatine Supplementation and Women Athletes

Joan Eckerson, PhD

1. INTRODUCTION

The effect of oral creatine (Cr) supplementation on performance has received a considerable amount of attention in recent years, and has been shown to be effective for enhancing both aerobic and anaerobic exercise performance. The majority of Cr research has used men as subjects; therefore, findings reported for men may not necessarily be generalized to women, and there has been some controversy regarding the effectiveness of Cr supplementation for improving high intensity exercise performance in women. For example, Forsberg et al. *(1)* reported that women have an approx 10% higher total muscle-Cr (TCr) content compared with men, suggesting that they may have a lower potential to Cr load and, therefore, may not experience improvements in performance similar to those observed for men. However, more recently, a number of studies have reported that there are no gender differences in the magnitude of change in phosphocreatine (PCr) and/or TCr content following Cr loading *(2,3)*, and that short-term supplementation is equally effective between genders for increasing indices of high intensity exercise performance *(4)*. The purpose of this chapter was to examine the morphological potential for women to respond to Cr supplementation, and its effects on exercise performance and body composition. Based on the findings, practical applications and recommendations for future study are discussed.

2. PHYSIOLOGICAL PROFILES OF CR RESPONDERS

Not all studies have found positive results on performance following Cr supplementation *(5–11)*. Conflicting findings between studies may be explained, in part, by methodological issues; however, some of

From: *Essentials of Creatine in Sports and Health*
Edited by: J. R. Stout, J. Antonio and D. Kalman © Humana Press Inc., Totowa, NJ

the disparity may also be explained by apparent differences in the physiological profiles of responders and nonresponders. Greenhaff et al. *(12)* have reported that approx 20–30% of individuals are nonresponders to Cr supplementation, which they defined as less than a 10 mmol/kg dry mass (dm) increase in resting TCr following 5 d of Cr loading (20 g/d), and suggested that it may be necessary to increase resting TCr stores by at least 20 mmol/kg dm to experience significant improvements in performance. Based on these guidelines *(12)*, Syrotuik and Bell *(13)* recently reported that responders to Cr supplementation (>20 mmol/kg dm in TCr) had lower initial levels of intramuscular free-Cr and -PCr, a greater percentage of type II muscle fibers, greater fiber type cross-sectional area, and a larger amount of fat-free mass (FFM) at preload when compared with nonresponders (<10 mmol/kg dm in TCr). In their study *(13)*, the loading dose was 0.3 g/d for 5 d and included 11 young men as subjects (mean age = 22.7 yr).

Although these findings may not be directly extrapolated to women, studies which have compared muscle morphology between genders *(14–16)* have shown that there are no significant differences in fiber type distribution (types I, IIa, and IIb), and that the differences in strength between men and women are largely because of the greater cross-sectional area of these fibers in men. Given that the distribution of fiber types is similar in men and women, it seems reasonable to suggest that women exhibiting the same physiological characteristics as the responders in the study by Syrotuik and Bell *(13)* would also be considered to be the best candidates for Cr supplementation. Further research is necessary to not only replicate the findings of Syrotuik and Bell *(13)* using a larger number of subjects, but to also directly compare the results with women.

3. CHANGES IN MUSCLE TCR AND PCR FOLLOWING CR LOADING

Reported increases in muscle-TCr content following Cr loading range from 9.5 to 27.6% *(2,3,17–20)* with most studies reporting an increase of approx 18–20% *(2,3,17–19)*, whereas increases in PCr have been reported to range from 6 to 17.8% *(3,17–19,21,22)*. A number of studies that have examined the effect of Cr loading on TCr and/or PCr content have used women as subjects *(2,3,21–23)*. In a study using 19 college-age women (19–22 yr), Vandenberghe et al. *(22)* reported that 4 d of Cr loading (20 g/d) increased ($p < 0.05$) muscle-PCr content by 6% compared with placebo (Pl), whereas Smith et al. *(21)* reported a 15%

increase in resting PCr content in five women and three men following 5 d of loading. In addition, Harris et al. *(2)* and McKenna et al. *(3)* reported mean increases in muscle-TCr of 17.2 and 18%, respectively, in pooled samples of men and women. In the study by McKenna et al. *(3)*, the authors reported that the muscle-TCr content before loading was not significantly ($p > 0.05$) different between men and women ($n = 8$, 120.9 ± 1.7 mmol/kg dm; $n = 6$, 125.3 ± 4.9 mmol/kg dm, respectively), and that there were no gender differences in the magnitude of change in TCr content following 5 d of Cr loading (men = 26 ± 6.6 mmol/kg dm [21.5% increase] and women = 23.8 ± 4.4 mmol/kg dm [19% increase], respectively). Harris et al. *(2)* also reported that there were no apparent effects of age or gender (n = five women and 12 men; age range = 20–62 yr) on muscle-TCr content following supplementation (TCr range =140–160 mmol/kg dm for all subjects). These results are in contrast to those of Forsberg et al. *(1)* who reported that women have an approx 10% higher muscle-TCr content compared with men. The study by McKenna et al. *(3)* was one of the first to directly examine potential gender differences as a result of Cr loading. More recently, Parise et al. *(23)* and Tarnopolsky and MacLennan *(4)* have also reported that men and women respond similarly to Cr loading.

Interestingly, the study by Parise et al. *(23)* did report a difference between genders for leucine kinetics following Cr supplementation. The specific purpose of the study was to examine the effect of Cr (20 g/d × 5 d → 5 g/d × 3–4 d) on indexes of protein metabolism in men ($n = 13$) and women ($n = 14$) to help determine the mechanism of action by which Cr induces increases in strength and FFM. As indicated above, supplementation resulted in similar increases in TCr (~13%) and PCr (~9%) for both genders, and there were no changes in body weight (BW) or FFM (FFM change = ~1 kg in men and 0.5 kg women, respectively) regardless of treatment. In addition, Cr was found to have no effect on protein synthesis. However, Cr did significantly ($p < 0.05$) reduce the plasma leucine rate of appearance (–7.5%; marker of protein breakdown) and leucine oxidation (–19.6%) in men, but not in women. These findings suggest that Cr does not increase protein synthesis but, rather, may have an anticatabolic effect in men.

Parise et al. *(23)* were unclear about exactly why the same results did not occur in women, as both men and women exhibited similar increase in TCr and PCr, but stated that the gender differences in leucine kinetics suggest that either the tissues responding to Cr are gender specific or the downstream signaling responses to a given Cr load was gender specific. Therefore, Parise et al. *(23)* recommended that more research be conducted to determine the mechanisms responsible for the alterations in protein metabolism following Cr supplementation in men, as well as the observed gender differences.

4. EFFECTS OF CR SUPPLEMENTATION ON STRENGTH PERFORMANCE

Because men and women respond similarly to Cr supplementation, it is also reasonable to assume that women would demonstrate similar gains in strength and muscular performance following supplementation to those demonstrated by men. Although research that has examined the effects of Cr on muscle strength using women is limited, supplementation appears to significantly enhance strength performance beyond training alone.

Vandenberghe et al. *(22)* were among the first to extensively investigate the effect of long-term Cr supplementation and resistance training on muscle performance in women. Subjects included 19 healthy, untrained women who ranged in age from 19 to 22 yr. Using a double-blind design, the subjects were matched for arm flexor strength and BW and assigned to either a Cr group ($n = 10$) or Pl group ($n = 9$). The Cr group was given a 20 g/d loading dose for 4 d (two, 2.5 g Cr tablets ingested four times daily, Novartis Nutrition, Berne, Switzerland), which was followed by a 10 wk maintenance phase in which they ingested one, 2.5 g tablet twice per day (5 g/d). The Pl group received maltodextrin tablets and followed the identical supplementation protocol as that of the Cr group.

During the 10 wk maintenance phase, all subjects performed variable resistance exercise (five sets, 12 repetitions at 70% RM for leg press, shoulder press, squat, leg extension, leg curl, and bench press) for 1 h three times per week. The one-repetition maximum (1RM) for each exercise was determined at the onset of the training program and at 5 wk and 10 wk. ^{31}P-nuclear magnetic resonance (NMR) spectroscopy of the right gastrocnemius muscle was used to determine muscle [PCr] before and after the loading dose, and at 5 and 10 wk of the resistance training program. Immediately following the NMR measurements, each subject completed an intermittent exercise test using their right arm on an isokinetic dynamometer to determine arm-flexor torque. After a 5-min warm-up, subjects performed five bouts of 30 dynamic maximal contractions of the arm flexors at 180°/s separated by 2-min rest. The average torque output generated across all five bouts was used as the representative torque value for statistical purposes. Body composition (through underwater weighing) was also determined at the onset of the study before the loading phase, and at 5 wk and 10 wk of the resistance training program.

At the end of the 10-wk training period, a subgroup of subjects ($n = 13$) continued for an additional 10 wk in which they did not train, but did ingest their respective supplement (Cr; $n = 7$ and Pl; $n = 6$). This was followed by a 4-wk period in which subjects stopped taking their supplements and continued to detrain. Muscle [PCr] through NMR of the gastrocnemius and isokinetic testing of the arm flexors was repeated after 3 and 10 wk of the detraining period, and at 1 and 4 wk following cessation of supplementation.

The results for [PCr] showed that the Cr group experienced a 6% increase ($p < 0.05$) in muscle [PCr] following the 4-d loading period compared with no change in the Pl group; and the increase after 5 wk and 10 wk of training was 7 and 10% of the baseline concentrations ($p < 0.05$), respectively. During the 10-wk detraining and supplementation period, muscle [PCr] was maintained at a higher level ($p < 0.05$) in the Cr group compared with the Pl group, and when supplementation ceased, [PCr] was still significantly greater ($p < 0.05$) in the Cr group at 1 wk; however, at 4 wk there were no significant differences between groups, which suggests that a 4-wk washout period was sufficient to return [PCr] to baseline levels.

There were no significant differences in arm flexor torque between groups following the loading phase; however, the Cr group exhibited significantly greater ($p < 0.05$) torque values at all other time-points until the end of the study at 24 wk when supplementation had ceased for 4 wk. Results for the resistance training program showed that both groups experienced significant increases in 1RM for each of the six exercises; however, the 1RM increases for leg press, leg extension, and squat following the 10-wk training program were 20–25% greater ($p < 0.05$) in the Cr group compared with Pl.

The results for body composition showed that the change in FFM was greater ($p < 0.05$) in the Cr group after both 5 wk (2 kg) and 10 wk (2.6 kg) of resistance training compared with Pl (1.1 kg and 1.6 kg, respectively). Although there was a trend for an increase in BW ($p = 0.12$) and a decrease in fat percent (fat [%]) ($p = 0.14$) for the Cr group, there were no significant differences between groups for either variable.

These findings indicated that long-term Cr supplementation was effective for increasing muscle strength and FFM above and beyond training alone in sedentary women, and that supplementation attenuated the decreases in strength that occurred during detraining. Vandenberghe et al. *(22)* speculated that Cr provided a greater stimulus for training, but what is of particular interest and is in contrast to several similar studies that used men as subjects *(24–27)* was that Cr supplementation markedly increased strength without significantly affecting BW or fat (%). This finding has also been reported by several others *(8,28,29,31–33)*, and suggests that women may benefit from the performance enhancing effects of Cr without experiencing considerable changes in BW.

In a related study, Brenner et al. *(28)* examined the effect of 5 wk of Cr supplementation on strength and body composition using 16 NCAA division I female lacrosse players (age range = 18–22 yr) during their preseason conditioning program. Using a double-blind randomized design, the athletes were assigned to either a Cr ($n = 7$) or Pl ($n = 9$) group. The Cr group received a loading dose of 20 g/d (4 × 5 g/d) in capsule form (SportPharma Inc., Concord, CA) for 7 d, followed by a maintenance dose of 2 g/d for the remainder of the study (24 d). The Pl group followed a similar dosing protocol, but ingested capsules filled with sucrose that were identical in appearance to those used for the Cr group. As part of their preseason conditioning program, all subjects completed a resistance training program three times per week that included free-weight bench press and leg extension using a Nautilus system. Body composition (through underwater weighing), 1RM for the bench press and leg extension, and a muscle endurance test (five sets of 30 repetitions of unilateral isokinetic knee extension at 180°/s separated by 1-min rest) were measured pre- and postsupplementation.

The results showed that the Cr group demonstrated a significantly greater ($p < 0.05$) increase in 1RM bench press compared with Pl (6.2 ± 2 kg vs 2.8 ± 1.8 kg, respectively); however, there were no significant differences between groups for FFM or fat (%), and both groups demonstrated a 0.05 kg increase ($p < 0.05$) in BW. These findings indicate that Cr supplementation for approx 1 mo increased upper body strength in trained athletes when compared with training alone. In agreement with Vandenberghe et al. *(22)*, the authors suggested that Cr may have provided a greater stimulus for training and, thereby, increased upper body strength.

Larson-Meyer et al. *(34)* examined the effect of Cr on strength and body composition following 13 wk of training in 14 female NCAA division I soccer players. In a double-blind manner, the athletes were randomly assigned to receive either Cr (2 × 7.5 g/d × 5 d → 5 g/d × 12 wk; $n = 7$; Sandco International, Tuscaloosa, AL) or a Pl ($n = 7$) dissolved in PowerAde® using identical dosing protocols. All subjects participated in the same progressive resistance training program for 13 wk. The subjects were tested for 1RM squat and bench press, vertical jump (VJ), and body composition (through dual energy X-ray absorptiometry [DEXA]) at baseline, 5 wk, and 13 wk.

The results showed that the Cr group experienced greater gains in 1RM strength compared with the Pl group for both the bench press (18% vs 9%, respectively) and the squat (24% vs 12%, respectively) following the 13 wk conditioning program. There were no significant differences between groups for VJ or body composition, and no significant changes in fat mass in either group. However, both groups demonstrated significant increases in BW (Cr = 2.5 kg; Pl = 3.6 kg) and FFM (Cr = 1.65 kg; Pl = +1.02 kg FFM) over the 13-wk training period, which indicates that the training was responsible for any observed increases in body composition.

Short-term Cr supplementation also appears to enhance strength performance in women with minimal effects on body composition. Kambis and Pizzedaz *(35)* examined the effect of 5 d of Cr loading (0.5 g/kg FFM divided into four equal does for 5 d) using 22 healthy, college-aged women ($X \pm$ SEM age = 20.3 ± 0.2 yr; BW = 61.3 ± 0.5 kg; fat (%) = 22.1 ± 0.9%; FFM = 47.6 ± 1.1 kg) on isokinetic strength of the preferred quadriceps group, thigh circumference, and BW. The women were matched for diet and exercise habits, phase of the menstrual cycle, and FFM. Subjects were assigned in a double-blind manner to receive either Cr (n = 11; phosphagen, EAS Inc., Golden, CO) or Pl (n = 11; white powder cornstarch) and were instructed to avoid resistance training and monitoring their BW during the study.

The results showed that time to peak torque for leg extension significantly decreased ($p < 0.05$), and that average power in leg extension and flexion significantly increased in the Cr group compared with Pl. However, there were no significant differences between groups for changes in BW, FFM, fat (%), midquadriceps circumference, or skinfold thickness of the measured thigh. Therefore, Cr significantly improved muscle performance without concomitant changes in BW or muscle volume.

The studies described earlier suggest that both long- and short-term Cr supplementation lead to increases in strength in both trained and untrained women without significantly affecting BW or fat (%). The findings for body composition are in contrast to those for men, as many studies report a 1–2% increase in BW following Cr loading *(36)*. In addition, a number of other studies that used pooled samples of men and women to examine the effects of Cr on anaerobic exercise performance *(30,31,37)* and substrate utilization *(23)* also reported that the effect of Cr on FFM and BW is less pronounced in women as compared with men.

5. EFFECT OF CR ON ANAEROBIC EXERCISE PERFORMANCE

A number of studies using women as subjects have examined the effects of Cr supplementation on other anaerobic indices of performance including anaerobic working capacity (AWC) *(29,30,38)*, VJ *(39)*, and repeated bouts of anaerobic cycling *(4,39,40)*, with the majority of studies showing favorable results. For example, Kirksey et al. *(39)* reported that 6 wk of Cr supplementation (0.30 g/kg/d) improved VJ, power output, and work capacity (through cycle ergometery testing), and FFM in a pooled sample of men and women collegiate track and field athletes (sprinters, jumpers, throwers) compared with Pl. In agreement, Tarnopolsky and MacLennan *(4)* showed that Cr (5 g, four times per day × 4 d; International Supplement Association, Hamilton, Ontario) increased peak and relative peak anaerobic cycling performance (3.7%), maximal voluntary contraction of the dorsiflexors (6.6%), and lactate (20.8%) with no gender-specific responses using a sample of 24 recreationally active men and women. The authors used a double-blind, Pl-controlled, crossover design, and it was concluded that both physically active men and women can increase indices of high-intensity exercise performance following short-term Cr supplementation.

In a related study, Eckerson et al. *(29)* examined the effect of 2 d and 5 d of Cr loading (20 g/d; Cr Edge Effervescent, FSI Nutrition Fortress Systems, LLC., Omaha, NE) on AWC and BW in 10 physically active women (mean age ± SD = 22 ± 5 yr) using a double-blind, crossover design. The authors reported that 5 d of supplementation resulted in a 22% increase in AWC ($p < 0.05$), whereas the Pl trial resulted in a 5% decline in performance ($p > 0.05$). The increase in AWC was comparable with values reported for men *(41,42)*, as well as combined samples of men and women (10–15%) *(38,43,44)*. Although there was a significant main effect for BW in the study by Eckerson et al. *(29)*, it was not because of Cr supplementation. The individual changes in BW from baseline to 5 d ranged from –0.45 kg to 0.91 kg and –0.45 kg to 1.36 kg for the Pl and Cr trials, respectively, and likely reflected daily fluctuations in BW.

In a follow-up study by Eckerson et al. *(30)* that compared the effects of 2 d and 5 d of supplementation with either Cr or Cr + phosphate salts on AWC in men and women, there were significant gender effects for AWC expressed both in absolute values (kJ) and relative to BW (kJ/kg); however, the women demonstrated no significant interactions for AWC or BW. The results for the men showed that Cr + phosphate significantly ($p < 0.05$) increased AWC following 2 d (23.8%) and 6 d (49.8%) of supplementation compared with Pl. Although the changes in AWC following 6 d of Cr or Cr + phosphate salts was not statistically significant for the women, AWC was increased by 13 and 10.8%, respectively, whereas the Pl group demonstrated a 1.1% decline. Therefore, these results may have some practical significance from a performance standpoint.

Ziegenfuss et al. *(40)* found that only 3 d of Cr supplementation (0.35 g/kg FFM) increased sprint cycle performance in NCAA division I athletes, and that the effect was greater in women (treatment groups pooled) as the sprints were repeated. The subjects included 10 men (eight wrestlers and two hockey players) and 10 women (three gymnasts, two basketball players, two field hockey players, two softball players, and one track athlete) who were matched for sex and 10-s cycle sprint scores, and paired by rank before being randomly assigned in a double-blind manner to receive either Cr ($n = 10$; American Biorganics, Aurora, OH) or Pl ($n = 10$; maltodextrin). Before and after supplementation, the subjects performed six, 10-s cycle sprints with 1 min recovery between bouts. In this study, the authors *(40)* chose a significance level of $p < 0.10$, as their opinion was that the relative risk of making a type II error was greater than that of making a type I error given the subject population (i.e., highly trained athletes). Therefore, the interpretation of the results is left to the discretion of the reader. Statistically significant increases were observed in the Cr group for BW (0.9 kg ± 0.1 kg; $p < 0.03$), total work during the first sprint ($p < 0.04$), and peak power during sprints 2–6 ($p < 0.10$) compared with the Pl group. Significant gender × sprint interactions revealed that during sprints 1 and 2, peak power and total work values were higher in the males ($p < 0.02$); but the reverse was true for the women during sprints 4–6 ($p < 0.01$). It is important to note that these results reflect the finding that women, in general, outperform men during fatiguing tasks *(45)* and did not account for any treatment effect of Cr *per se*. However, the results of this study are intriguing as it showed that highly trained athletes might experience increases in anaerobic exercise performance after only 3 d of supplementation.

6. THE EFFECT OF CR ON SPORTS PERFORMANCE

Although there is mounting evidence that indicates that women can increase strength and other various indices of anaerobic exercise performance as a result of Cr supplementation, fewer studies have taken the next step to directly determine how Cr may influence win–loss records and the execution of skills during competition. For example, because Cr may provide a greater stimulus for training by improving the recovery time between successive bouts of training, it seems reasonable to assume that this will carry-over to competition and allow athletes to recover faster and perform better during sporting events that may occur on the same day (i.e., softball double-header, tennis matches, swim competition, and track and field events) or on successive days of competition (i.e., conference basketball, soccer, and volleyball tournaments). However, few studies have bridged this gap, which makes the area ripe for research.

Cox et al. *(46)* investigated the effects of short-term Cr supplementation (4 × 5 g/d × 6 d) on performance during a field test simulating soccer match play using elite female soccer players from the Australian National Team and found that the Cr group (n = 6) experienced significant increases ($p < 0.05$) in BW (0.8 kg) and significantly improved repeated sprint performance and some agility tasks that mimicked soccer play compared with the Pl group (BW gain = 0.3 kg). However, Cr had no effect on shooting accuracy. These findings are in agreement with Ziegenfuss et al. *(40)* who also reported that elite athletes can benefit from short-term Cr supplementation; however, the subjects in that study did not perform exercise tasks that mimicked their respective sport.

One sport that does allow investigators to easily mimic competition is swimming. Several studies *(10,11,32,47–49)* using a wide range of highly trained swimmers have generally found that Cr is ineffective for increasing swim sprint performance. The athletes used in these studies included junior swimmers (~16 yr) *(47,48)*, collegiate swim teams *(11,32)*, as well as national and international caliber athletes *(10)*. Most of the studies used pooled samples of men and women *(10,32,47,48)* and used distances ranging from 25 *(10)* to 400 m *(11)*, with 50 and 100 m being used most frequently *(10,11,32,47–49)*.

With the exception of one study by Grindstaff et al. *(48)*, most research has shown that Cr supplementation is ineffective for improving single sprint swim performance *(10,11,32,47,49)*. However, when the athletes were required to perform repeated swim sprints, some studies *(48,49)* have reported improvements in the time to complete the series, as well as reported increases in work and power output *(48)*. Perhaps the lack of improvement in the majority of studies that used single swim sprints is because of the fact that these performance tests did not require recovery between several bouts of exercise and, therefore, did not accurately represent the potential efficacy of Cr supplementation. One of the benefits of increasing Cr stores is that it may reduce neuromuscular fatigue and delay the onset of lactate accumulation during high-intensity exercise. Therefore, in the studies in which the athletes performed a single, short-duration sprint before and after supplementation, fatigue was not a major issue; whereas in the studies that required repeated sprints, the subjects may have more fully realized the benefits of Cr on fatigue resistance *(50)* and, as a result, the decline in performance over time was attenuated. Although it is difficult to control for the various biological and environmental factors that affect performance on any given day and during competition, future research is warranted to more fully explain how Cr supplementation directly affects outcomes on the playing field.

7. CR AND AEROBIC EXERCISE

Although Cr is best suited for athletes involved in activities that require repeated bouts of high-intensity exercise, there is some evidence to suggest that Cr might enhance oxygen uptake and recovery from aerobic exercise. Studies using females only are sparse; however, in a study by Aoki et al. *(51)* the authors reported that women who supplemented with Cr for 12 d were able to complete more repetitions of leg-press at 80% 1RM following a 20 min run compared with control subjects. The specific aim of the study was to determine how Cr would effect resistance exercise performance (1RM and three sets to fatigue at 80% 1RM) following a 20-min run in which subjects were instructed to run as far as they could during the allotted time. There were no significant differences between the two groups for the distance covered during the 20 min run or 1RM testing following the run; however, there was a significant decline in the number of maximal repetitions performed during the last two sets in the Pl group vs Cr group. The authors *(51)* suggested that Cr may have enhanced recovery, which allowed subjects to perform better during the resistance exercise following the run.

Nelson et al. *(52)* found that Cr allowed subjects to perform submaximal workloads at a lower oxygen cost (VO_2), and reduced the work performed by the cardiovascular system in a study which determined the effects of Cr on cardiorespiratory responses during a graded exercise test (GXT). The subjects were 36 trained men ($n = 20$) and women ($n = 16$) (age range = 21–27 yr) who were randomly placed into either a Cr group (n = 13 men, six women; 20 g/d × 7 d) or a Pl group (n = seven men, 10 women) and performed a GXT using a Monark cycle ergometer pre- and post-treatment. The results showed that Cr significantly increased ($p < 0.05$) total test time (20.3 ± 4 min to 21.5 ± 3.5 min) compared with Pl (17.3 ± 3 min to 17.4 min ± 3 min), and that VO_2 and heart rate at the end of first five GXT stages were significantly lower for Cr vs no change for Pl. In addition, the ventilatory threshold increased significantly from pre- to post-testing for the Cr group (66 to 78% peak VO_2, respectively), whereas the Pl group demonstrated no change (70 to 68%

peak VO_2, respectively). The authors *(52)* speculated that the decreases in submaximal VO_2 and heart rate were because of increased stores of PCr in the muscle, which may have delayed the decrease in the ATP/ADP ratio and, in turn, delayed mitochondrial respiration and lowered VO_2.

In a related study that used female NCAA division I athletes (19 ± 2 yr), Stout et al. *(33)* examined the effect of Cr loading (4 × 5 g/d × 5 d; Cr Edge Effervescent) on the onset of the neuromuscular fatigue threshold by monitoring electromyographic fatigue curves from the vastus lateralis muscle using the physical working capacity at the fatigue threshold (PWC_{FT}) test. The PWC_{FT} represents the highest power output that results in a nonsignificant ($p > 0.05$) increase in the electrical activity of the thigh muscles over time. The results showed that the postsupplementation PWC_{FT} value for the Cr group ($n = 7$; $X = 186$ *W*) was significantly higher than that of the Pl group ($n = 8$; $X = 155$ *W*). The authors *(33)* suggested that the delay in neuromuscular fatigue may have been due to the effect of increased muscle-PCr content on the transition from aerobic to anaerobic metabolism and, therefore, Cr may have reduced the reliance on anaerobic metabolism and attenuated the accumulation of lactate and ammonia in the working muscles and blood.

Although more research is warranted to better understand the effects of Cr supplementation on aerobic forms of exercise, as well as recovery following aerobic exercise, the results of the studies mentioned above *(33,51,52)* tend to suggest that female (and male) endurance athletes may benefit from Cr supplementation by decreasing the work on the cardiovascular system during training at submaximal intensities and buffering the lactate response to exercise, which may allow for enhanced recovery during successive days of training. Refer to Chapter 3 for more in depth review on the effect of Cr on endurance performance.

8. PREVALENCE OF CR USE BY WOMEN ATHLETES

Although there is a preponderance of evidence to indicate that women athletes have the potential to enhance their performance beyond training alone without experiencing considerable increases in BW and fat (%), Cr is not a widely used dietary supplement by this population. Labotz and Smith *(53)* examined the prevalence and pattern of Cr use among 749 NCAA division I varsity and junior varsity athletes at the University of North Carolina through an anonymous survey. The results showed that 85% of the men had heard of Cr and that 48% had reported using the supplement. In contrast, only 4% of women athletes reported using Cr, with 38% reporting that they had at least heard of the supplement. When athletes were asked to describe their loading and maintenance dosages, more than 42% left the question unanswered or responded that they did not know. The most common method of quantifying their dosage amount was by reporting the number of "scoops" of Cr consumed.

In a related study, Greenwood et al. *(54)* determined the patterns of dietary Cr use in NCAA division I athletes (n = 63 women; n = 156 men) at a midsouth university, and also assessed their perceptions regarding Cr dosages and supplementation schedules. The results of the questionnaire showed that 88 male athletes reported using Cr as a dietary supplement, which represented 56% of the sample, whereas only two women (3%), a tennis player and a track and field athlete, reported using Cr. Reasons cited by the women athletes for not using Cr included their uncertainty of its safety and associated side effects. The Cr users reporting positive effects (n = 80) were typically at or below recommended dosages for the loading phase ($X \pm$ SD = 0.30 ± 0.24 g/kg ingested for 6.6 ± 10.2 d), but were ingesting more than the recommended dosage during the maintenance phase (0.15 ± 0.14 g/kg ingested for 101 ± 195 d), and that those who reported no effects (n = 10) ingested loading dosages that were below the recommended amount, but above the dosage recommended for the maintenance phase. These findings *(54)*, as well as those of Labotz and Smith *(53)* indicate that there is a need for education among college athletes regarding Cr supplementation dosing and scheduling, particularly among women athletes.

It is likely that women avoid Cr because of a concern about undesirable side effects such as weight gain, bloating, cramping, and diarrhea. However, short-term side effects are widely exaggerated and gastrointestinal problems are typically associated with dosing or improper mixing *(55)*. In fact, Mihic et al. *(37)* examined the effect of acute Cr supplementation (20 g/d × 5 d) on FFM and potential side effects in physically active men (n = 15) and women (n = 15) and found no effects of Cr on blood pressure, creatinine clearance, or plasma Cr-kinase activity. However, there was a significant gender effect for Cr-induced increases in BW and a nonsignificant trend (p = 0.052) for increases in FFM. The average increase in BW and FFM for the men was 1.6 kg and 1.4 kg, respectively, whereas the women experienced negligible increases (BW = 0.45 kg and FFM = 0.44 kg) that were not significantly different from the women in the Pl group. These findings are in agreement with several others *(8,28–33)* and bears repeating that Cr appears to have a greater effect on BW and FFM in men than women. Therefore, the concern by some women athletes that their BW and/or fat (%) will increase as a result of Cr supplementation is largely unwarranted.

Ms. Jennifer Yee, M.A., CSCS, NSCA-CPT, Assistant Strength Coach, Creighton University 2002 – 2006; Personal Trainer

Q: What is the prevalence of Cr use by NCAA division I female athletes with whom you have worked with?

A: Cr is not used by a majority of our athletes, and I can recall only two female athletes that have asked about it, a soccer player and a basketball player. The soccer player was genetically inclined to build muscle and was interested in increasing her muscle mass through training and supplementation because her coach suggested that she move from a forward position to a defender position. Over a summer, she increased her training and took Cr and gained approx 8–10 lbs of muscle and decreased her body fat by 1–2%. She claimed that she recovered faster between her summer workouts and that she was able to lift more weight than she ever had before by the end of the summer. She also felt that she had a much easier time maintaining the added muscle mass over the season. The basketball player was very lean and skinny and her coach suggested that she try Cr to add some mass. Although most women don't experience dramatic increases in muscle mass as a result of Cr use, she was not interested in taking a supplement that she perceived would lead to weight gain, even if it was muscle. As I mentioned before, my soccer player had the genetics to put on muscle, therefore, I believe her results were the exception and not the norm.

Q: Why don't many women athletes even consider experimenting with Cr?

A: I think that since a lot of athletes use Cr to help gain muscle and weight, women stay away from it. Even though most female athletes are too thin, they still believe that thinner is better! Also, a lot of athletes (both men and women) don't want to spend money on Cr. If we could give it to them for free, like we used to be able to do before the NCAA rules changed, I think the use would be more prevalent. As it stands now, men are more likely to spend money on supplements, but not women.

Q: Are you allowed by the NCAA to monitor the athlete's Cr use, even though you can't distribute the supplement?

A: Yes, we can monitor their use by helping them with doses and measurements, but we can't give it to them. In fact, it doesn't make much sense to me that we can suggest that they use Cr and can tell them where to buy it, but we can't provide it to them ourselves.

Q: Would you recommend Cr to your female athletes and, if so, what would be the circumstances with which you would recommend it?

A: Yes, I would recommend it based on the research that I have read, but first I would examine their current dietary habits to determine if there are any deficiencies that need to be addressed. If they are eating well and staying hydrated, I feel comfortable recommending Cr to my athletes who are trying to increase muscle mass and recover faster between workouts (i.e., two-a-days, lifting hard four to six times per week, and so on).

9. EFFECTS OF THE MENSTRUAL CYCLE

No studies have directly examined the effect of Cr uptake into the muscle during different phases of the menstrual cycle, or whether menstrual dysfunction as a result of energy restriction (amenorrhea, oligomenorrhea, delayed menarche, and so on) effects Cr uptake. What is, perhaps, a bigger concern at the present time is the question of whether or not there is a need to control for the menstrual cycle during intervention trials using Cr, because this may be considered a confounding factor regarding the changes in BW and performance. Tarnopolsky *(56)* recommends testing women during the follicular phase to control for hormonal imbalances that may affect BW and other menstrual symptoms; however, female athletes must compete and train during all phases of their menstrual cycle and several studies suggest that the phases of the ovarian cycle have no effect on high intensity, anaerobic exercise performance.

Giacomoni et al. *(57)* examined the effect of the menstrual cycle phase and menstrual symptoms on maximal anaerobic performance in seven eumenorrheic women and 10 women using oral contraceptives (mean ± SD age = 23 ± 3 yr). The subjects performed three anaerobic tests to determine maximal cycling power, jump power, and jump height during menses (days 1–4), the midfollicular phase (days 7–9), and the midluteal phase (days 19–21) of the ovarian cycle. The results showed that the menstrual cycle phase had no effect ($p > 0.05$) on maximal anaerobic jumping and cycling performance, and this was true for both the eumenorrheic subjects, as well as the subjects using oral contraceptives. These findings are in agreement with others who have reported no effect of menstrual cycle phase on anaerobic exercise *(58–60)*, as well as cycling time at trial performance *(61)*.

Wehnert-Roti and Clarkson *(62)* examined the effect of Cr supplementation on BW during the luteal and follicular phases of the menstrual cycle in nine eumenorrheic women (X ± SE age = 26.7 ± 1.7 yr) using a repeated measures design and reported no significant ($p > 0.05$) differences in BW gain between the two phases (luteal phase = +0.25 kg and follicular phase = +0.40 kg). However, they did report that there was considerable intersubject variability in the response, and that some subjects maintained or lost weight during the two phases. An interesting finding in their study *(62)* showed that women who ate more protein (15–59%) as part of their daily caloric intake demonstrated no change or decreases in BW, whereas subjects who ate less protein daily (11–13% of total intake) gained weight. The correlation between percent dietary protein intake and the change in BW was significant ($r = -0.57$, $p < 0.05$), which prompted the authors to suggest that the increases in BW demonstrated by some subjects following Cr supplementation may be influenced by the composition of the diet.

Research testing during the follicular phase may control for the potential adverse effects of premenstrual syndrome (bloating, weight gain, and cramping) that some athletes experience; however, for the coach or health practitioner, it seems reasonable to suggest that athletes could begin Cr supplementation during any part of their menstrual cycle. Further research may be warranted to determine the effects, if any, of Cr uptake into the muscle during different phases of the menstrual cycle.

10. CONCLUSION

A considerable number of studies (Table 1) are showing mounting evidence that women appear to have the same potential to benefit from Cr supplementation as men. Interestingly, however, very few female athletes report using Cr as a dietary supplement to enhance their performance. Although there may be a large amount of variability in the response to Cr by individual female athletes, based on the available research, it has been shown to be a safe and effective ergogenic aid that does not markedly affect BW, FFM, or fat (%) in women. Therefore, the reluctance among some women athletes to experiment with Cr because of a fear of weight gain or other adverse side effects may be unfounded. Furthermore, many adverse events reported anecdotally by individuals that use Cr do not tend to be substantiated in controlled research studies and, therefore, are likely because of improper mixing and/or dosing. Survey research does, indeed, indicate that athletes need to be educated about Cr dosing and scheduling, and should be monitored closely by their coaches.

11. PRACTICAL APPLICATIONS

A summary of the information presented earlier indicates that:

- The changes in [TCr] and [PCr] in women following Cr supplementation are similar to those observed for men.
- Cr results in increases in strength that are similar in magnitude to the changes demonstrated by men, and that these increases have been demonstrated in both trained and untrained women.

Table 1
Summary of Cr Studies on Exercise Performance and Body Composition in Women

Subjects	*Dosage*	*Findings*	*References*
Strength performance			
College-age women Cr = 11; Pl = 11	0.5 g/kgFFM × 5 d	↑ peak Tq; ↔ between groups for BW, FFM, fat (%)	*35*
Female collegiate soccer Cr = 7; Pl = 7	2 × 7.5 g/d × 5 d → 5 g/d × 12 wk	↑ 1RM BP and squat; ↔ between groups for BW, FFM, FM, VJ	*34*
Female collegiate lacrosse Cr = 7; Pl = 9	4 × 5 g/d × 7 d → 2 g/d × 24 d	↑ 1RM BP; ↔ between groups for BW, FFM, fat (%)	*28*
Sedentary college-age women Cr = 10; Pl = 9	4 × 5 g/d × 4 d → 5 g/d × 10 wk (training) → 5 g/d × 10 wk (no training) → 0 g/d × 4 wk	↑ [PCr], ↑ arm flexor Tq, ↑ 1RM leg press, leg extension, and squat, ↑ FFM; ↔ between groups for BW, fat (%)	*22*
Anaerobic exercise performance			
Physically active college-age women; CrP $n = 10$; Cr = 10; Pl = 10 (31 men also studied)	4 × 5 g/d × 6 d	Cr = 13% ↑, CrP = 11% ↑, Pl = 1.1% ↓ in AWC, but N.S. between groups; ↔ for BW between groups; gender effects found	*30*

(Continued)

Table 1 *(Continued)*

Subjects	*Dosage*	*Findings*	*References*
Physically active college-age women; $n = 10$; cross-over design (Cr and Pl)	4×5 g/d $\times$ 5 d	↑ AWC 22%; ↔ BW between treatment	*29*
Men and women collegiate athletes; Cr = 10; Pl = 10	0.35 g/kg FFM $\times$ 3 d	↑ total work and peak power during repeated sprint cycling; ↑ BW	*40*
Physically active college-age men ($n = 12$) and women ($n = 12$); cross-over design (Cr and Pl)	4×5 g/d $\times$ 4 d	↑ peak and relative peak anaerobic cycling power, ↑ dorsi-flexion MVC Tq, ↑ [La]; no gender effects found	*4*
Men ($n = 7$) and women ($n = 8$) collegiate soccer players; cross-over design (Cr and Pl)	0.3 g/kg BW $\times$ 6 d	↔ treadmill running with sprint intervals; ↑ BW for men only	*5*
Men ($n = 16$) and women ($n = 20$) collegiate track and field athletes; Cr = 15; Pl = 21	22 g/d $\times$ 6 wk	↑ VJ, ↑ mean cycle peak power, mean power output, and total work, ↑ FFM	*39*
Physically active college-age men (n =8) and women ($n = 7$); Cr = 7; Pl = 8	4×5 g/d $\times$ 5 d	↑ AWC 10%; ↔ critical power	*38*

Sports performance			
Men ($n = 8$) and women ($n = 7$) NCAA division III swimmers; Cr and Pl (group n not provided)	0.3 g/kg BW × 5 d → 2.25 g/d × 9 d	↑ 100 yd single sprint; ↔ 50 yd sprint or BW between groups	*32*
Elite women soccer players Cr = 6; Pl = 6	4 × 5 g/d × 6 d	↑ sprint and agility, ↑ BW; ↔ shooting accuracy	*46*
Men ($n = 10$) and women ($n = 10$) competitive junior (~16 yr) swimmers Cr = 10; Pl = 10	4 × 5 g/d × 5 d → 5 g/d × 22 d	↔ 50 and 100 m single sprint between groups	*47*
Men (18) and women (14) collegiate swimmers; Cr (n = seven women, nine men); Pl (n = seven women, nine men)	4 × 5 g/d × 6 d → 2 × 5 g/d × 6 d	Gender effects found; ↔ swim velocity between groups for women; ↑ swim velocity for men (6 × 50 m intervals); ↔ BW, FFM, fat (%) for men and women	*8*
Men ($n = 7$) and women ($n = 11$) competitive junior swimmers (~15 yr); Cr (n = six women, three men); Pl (n = five women, four men)	3 × 7 g/d × 9 d	↑ swim times for repeated 50 m and 100 m sprints; ↔ BW, FFM, FM, fat (%), TBW between groups	*48*

(Continued)

Table 1 *(Continued)*

Subjects	*Dosage*	*Findings*	*References*
Women collegiate swimmers (n =10); Cr and Pl (group n not provided)	2 g/d × 6 wk	↔ 100 m or 400 m single sprint time; ↔ FFM; ↔ PCr resynthesis or mitochondrial ATP synthesis	*11*
Men (n = 11) and women (n =9) national and international competitive swimmers; Cr = 10; Pl = 10	4 × 5 g/d × 5 d	↔ 25 m, 50 m, or 100 m single sprint times; ↔ [La]; ↑ BW	*10*
Aerobic exercise performance			
College-age women (n = 14); Cr and Pl (group n not reported)	4 × 5 g/d × 5 d → 3 g/d × 7 d	↑ leg press reps (80% 1RM) following a 20 min aerobic run; ↔ 1RM between groups following the run	*51*
Physically active men (n = 20) and women (n = 16); Cr (n = six women, 13 men); Pl = 10 women, seven men)	4 × 5 g/d × 7 d	↑ GXT test time; ↑ VT; ↓ submax VO_2; ↓ HR	*52*
Women collegiate athletes Cr = 7; Pl = 8	4 × 5 g/d × 5 d	↑ neuromuscular fatigue threshold through PWC_{FT} test; ↔ BW between groups	*33*

Cr, creatine; Pl, placebo; 1RM, one-repetition maximum; BW, body weight; FFM, fat-free mass; FM, fat mass; fat (%), percentange of body fat; N. S., not significant; Tq, torque; PCr, phosphocreatine; VJ, vertical jump; La, lactate; GXT, graded exercise test; VT, ventilatory threshold; HR, heart rate; PWC_{FT}, physical working capacity at fatigue threshold; ↑ indicates increase; ↓ indicates decrease; ↔ indicates no change.

- Cr increases anaerobic performance and may especially benefit women athletes involved in sports that require repeated short bursts of high-intensity exercise (i.e., basketball, volleyball, sprinters and throwers, soccer, field hockey, and tennis).
- Women exhibit smaller increases in BW and FFM vs men following short- and long-term Cr supplementation. Therefore, women athletes may be able to enhance their performance without the "adverse" side effect of a significant increase in BW.
- Cr might enhance aerobic metabolism and improve recovery between successive days of training.
- Cr is *not* an anabolic steroid—it enhances performance by reducing fatigue, which allows for a greater stimulus of training.
- Cr is safe and effective when taken in recommended doses (but would be contraindicated in individuals with known kidney or liver problems).
- Cr does not appear to exacerbate premenstrual symptoms.

12. DOSING STRATEGIES FOR WOMEN

It is important to keep in mind that there may be wide variability in the individual response of women athletes to Cr supplementation. It is also reasonable to suggest that women athletes who exhibit the same physiological characteristics as the responders identified in the study by Syrotuik and Bell *(13)* would also be considered to be the best candidates for Cr supplementation.

Although it is typically recommended that individuals Cr load for 4–7 d, Harris et al. *(2)* reported that Cr uptake into muscle is greatest during the first 2 d of loading (6 × 5 g/d × 7 d), with approx 20% of the Cr taken up as PCr. Hultman et al. *(19)* have also reported that a dose of 3 g/d for 28 d is as effective as Cr loading for increasing muscle-TCr. Therefore, "slowly loading" the muscle with Cr may result in significant increases in performance and alleviate any real or perceived side effects that are sometimes associated with 5–7 d of loading. Based on this information, the following represents two dosing strategies recommended for women athletes:

12.1. Protocol I—(Jeff Stout, Ph.D.; personal communication June 2007)

- 3–6 g/d for 6–8 wk.
- Ingest within 20–30 min postexercise with CHO (50 g or 1 g/kg).
- If taken with CHO and protein, use a 2:1 or 3:1 ratio of CHO to protein.
- Cr Monohydrate most commonly studied.
- 5–6 wk washout.

Example for a 140 lb (63.5 kg) female athlete using a 2:1 ratio of CHO:protein.

Ingest 0.8 g/kg CHO + 0.4 g/kg PRO + 5 g Cr within 20–30 min postexercise = 51 g CHO + 25 g PRO + 5 g Cr.

12.2. Protocol II—Recommended by Mesa et al. (36)

- Day 1: 4 × 5 g Cr with 500 mL containing 90–100 g simple CHO.
- Day 2: 4 × 5 g Cr with 47 g simple CHO and 50 g protein.
- Day 3: start maintenance phase; 3–5 g/d with simple CHO.
- Ingest three to four times per week vs daily to maintain TCr and prolong washout effects.
- Ingest the CHO 30 min after Cr ingestion to produce peak Cr and insulin concentrations at similar time-points.

It is imperative that women athletes and their coaches keep in mind that Cr is a dietary supplement and, therefore, is *not* a substitute for a sound nutrition and training program. Female athletes, in particular, should first be encouraged to consume an adequate number of total kcals, including good quality protein, before considering Cr supplementation. At Ball State University, Pearson *(63)* reported four common threads that surfaced when the diets of university athletes were examined: (1) they did not consume enough total kcals, (2) they skipped breakfast, (3) did not consume enough protein, and (4) consumed high amounts of caffeine. Although this was not a scientifically based study, it most likely rings true for a large number of athletes. Therefore, when an athlete inquires about Cr supplementation, whether male or female, it is recommended that coaches and team sports nutritionists first determine that they are already eating well, training hard, getting enough sleep, and staying hydrated. If Cr supplementation is advised, it is also imperative that the dosing strategy used is closely monitored, because many athletes are unfamiliar with appropriate dosing protocols and often rely on untrustworthy sources of information regarding Cr dosing strategies.

REFERENCES

1. Forsberg AM, Nilsson E, Werneman J, Bergstrom J, Hultman E. Muscle composition in relation to age and sex. Clin Sci 1991; 81:249–256.
2. Harris RC, Soderlund K, Hultman E. Elevation of creatine in resting an exercised muscle of normal subjects by creatine supplementation. Clin Sci 1992; 83:367–374.
3. McKenna MJ, Morton J, Selig SE, Snow RJ. Creatine supplementation increases muscle total creatine but not maximal intermittent exercise performance. J Appl Physiol 1999; 87(6):2244–2252.

4. Tarnopolsky MA, MacLennan DP. Creatine monohydrate supplementation enhances high-intensity exercise performance in males and females. Intl J Sport Nutr Exerc Metab 2000; 10:452–463.
5. Biwer CJ, Jensen RL, Schmidt WD, Watts PB. The effect of creatine on treadmill running with high-intensity intervals. J Strength Cond Res 2003; 17(3):439–445.
6. Febbraio MA, Flanagan TR, Snow RJ, Zhao S, Carey MF. Effect of creatine supplementation on intramuscular TCr, metabolism and performance during intermittent, supramaximal exercise in humans. Acta Physiol Scand 1995; 155:387–395.
7. Finn JP, Ebert TR, Withers RT, et al. Effect of creatine supplementation on metabolism and performance in humans during intermittent sprint cycling. Eur J Appl Physiol 2001; 84:238–243.
8. Leenders N, Sherman WM, Lamb DR, Nelson TE. Creatine supplementation and swimming performance. Int J Sport Nutr 1999; 9:251–262.
9. Misic M, Kelley GA. The impact of creatine supplementation on anaerobic performance: A meta-analysis. Am J Med Sports 2002; 4:116–124.
10. Mujika I, Chatard JC, Lacoste L, Barale F, Geyssant A. Creatine supplementation does not improve sprint performance in competitive swimmers. Med Sci Sports Exerc 1996; 28(11):1435–1441.
11. Thompson CH, Kemp GJ, Sanderson AL, et al. Effect of creatine on aerobic and anaerobic metabolism in skeletal muscle in swimmers. Br J Sports Med 1996; 30:222–225.
12. Greenhaff PL, Bodin K, Soderlund K, Hultman E. Effect of oral creatine supplementation on skeletal muscle phosphocreatine resynthesis. Am J Physiol 1994; 266:E725–E730.
13. Syrotuik DG, Bell GJ. Acute creatine monohydrate supplementation: a descriptive physiological profile of responders vs. nonresponders. J Strength Cond Res 2004; 18(3):610–617.
14. Miller AEJ, MacDougall JD, Tarnopolsky MA, Sale DG. Gender differences in strength and muscle fiber characteristics. Eur J Appl Physiol 1993; 66:254–262.
15. Staron RS, Hagerman FC, Hikida RS, et al. Fiber type composition of the vastus lateralis muscle of young men and women. J Histochem Cytochem 2000; 48(5):623–629.
16. Toft I, Lindal S, Bonaa KH, Jenssen T. Quantitative measurement of muscle fiber composition in a normal population. Muscle Nerve 2003; 28:101–108.
17. Balsom PD, Soderlund K, Sjodin B, Ekblom B. Skeletal muscle metabolism during short duration high-intensity exercise: influence of creatine supplementation. Acta Physiol Scand 1995; 154:303–310.
18. Casey A, Constantin-Teodosiu D, Howell S, Hultman E, Greenhaff PL. Creatine ingestion favorably affects performance and muscle metabolism during maximal exercise in humans. Am J Physiol 1996; 271:E31–E37.
19. Hultman E, Soderlund K, Timmons JA, Cederblad G, Greenhaff P. Muscle creatine loading in men. J Appl Physiol 1996; 81:232–237.
20. Nelson AG, Arnall DA, Kokkonen J, Day R, Evans J. Muscle glycogen supercompensation is enhanced by prior creatine supplementation. Med Sci Sports Exerc 2001; 33(7):1096–1100.
21. Smith S, Montain SJ, Matott RP, Zientara GP, Jolesz FA, Fielding RA. Effects of creatine supplementation on the energy cost of muscle contraction: a ^{31}P-MRS study. J Appl Physiol 1999; 87(1):116–123.

22. Vandenberghe K, Goris M, Van Hecke P, Van Leemputte M, Vangerven L, Hespel P. Long-term creatine intake is beneficial to muscle performance during resistance training. J Appl Physiol 1997; 83(6):2055–2063.
23. Parise G, Mihic S, MacLennan D, Yarasheski KE, Tarnopolsky MA. Effects of acute creatine monohydrate supplementation on leucine kinetics and mixed muscle protein synthesis. J Appl Physiol 2001; 91:1041–1047.
24. Becque MD, Lochmann JD, Melrose DR. Effects of oral creatine supplementation on muscular strength and body composition. Med Sci Sports Exerc 2000; 32(3):654–658.
25. Maganaris CN, Maughan RJ. Creatine supplementation enhances maximum voluntary isometric force and endurance capacity in resistance trained men. Acta Physiol Scand 1998; 163:279–287.
26. Peeters BM, Lantz CD, Mayhew JL. Effect of oral creatine monohydrate and creatine phosphate supplemenatation on maximal strength indices, body composition, and blood pressure. J Strength Cond Res 1999; 13(1):3–9.
27. Volek JS, Duncan ND, Mazzetti SA, et al. Performance and muscle fiber adaptations to creatine supplementation and heavy resistance training. Med Sci Sports Exerc 1999; 31(8):1147–1156.
28. Brenner M, Walberg-Rankin J, Sebolt D. The effect of creatine supplementation during resistance training in women. J Strength Cond Res 2000; 14(2):207–213.
29. Eckerson JM, Stout JR, Moore GA, Stone NJ, Nishimura K, Tamura K. Effect of two and five days of creatine loading in anaerobic working capacity in women. J. Strength Cond Res 2004; 18(1):168–173.
30. Eckerson JM, Stout JR, Moore GA, et al. Effect of creatine phosphate supplementation on anaerobic working capacity and body weight after two and six days of loading in men and women. J Strength Cond Res 2005; 19(4):756–763.
31. Lehmkuhl M, Malone M, Justice B, et al. The effects of 8 weeks of creatine monohydrate and glutamine supplementation on body composition and performance measures. J Strength Cond Res 2003; 17(3):425–438.
32. Selsby JT, Beckett KD, Kern M, Devor ST. Swim performance following creatine supplementation in Division III athletes. J Strength Cond Res 2003; 17(3):421–423.
33. Stout J, Eckerson J, Ebersole K, et al. Effect of creatine loading on neuromuscular fatigue threshold. J Appl Physiol 2000; 88:109–112.
34. Larson-Meyer DE, Hunter GR, Trowbridge CA, et al. The effect of creatine supplementation on muscle strength and body composition during off-season training in female soccer players. J Strength Cond Res 2000; 14(4):434–442.
35. Kambis KW, Pizzedaz SK. Short-term creatine supplementation improves maximum quadriceps contraction in women. Int J Sport Nutr Exerc Metab 2003; 13:87–96.
36. Mesa JLM, Ruiz JR, Gonzalez-Gross MM, Gutierrez Sainz A, Castillo Garzon MJ. Oral creatine supplementation and skeletal muscle metabolism in physical exercise. Sports Med 2002; 32(14):903–944.
37. Mihic S, MacDonald JR, McKenzie S, Tarnopolsky MA. Acute creatine loading increases fat-free mass, but does not affect blood pressure, and plasma creatinine, or CK activity in men and women. Med Sci Sports Execr 2000; 32(2):291–296.
38. Smith JC, Stephens DP, Hall EL, Jackson AW, Earnest CP. Effect of oral creatine ingestion on parameters of the work rate-time relationship and time to exhaustion in high intensity cycling. Eur J Appl Physiol 1998; 77:360–365.

39. Kirksey B, Stone MH, Warren BJ, et al. The effects of creatine monohydrate supplementation on performance measures and body composition in collegiate track and field athletes. J Strength Cond Res 1999; 13(2):148–156.
40. Ziegenfuss TN, Rogers M, Lowery L, et al. Effect of creatine loading on anaerobic performance and skeletal muscle volume in NCAA Division I athletes. Nutrition 2002; 18:397–402.
41. Miura A, Fumiko K, Kajitani S, Sato H, Fukuba Y. The effect of oral creatine supplementation on the curvature constant parameter of the power-duration curve for cycle ergometry in humans. Jpn J Physiol 1999; 49:169–174.
42. Stout JR, Eckerson JM, Housh TJ, Ebersole KT. The effects of creatine supplementation on anaerobic working capacity. J Strength Cond Res 1999; 13(2):135–138.
43. Earnest CP, Stephens DP, Smith JC. Creatine ingestion effects time to exhaustion during estimation of the work rate-time relationship [Abstract]. Med Sci Sports Exerc 1997; 29:S285.
44. Hall EL, Smith JC, Stephens DP, Snell PG, Earnest CP. Effect of oral ingestion of creatine monohydrate on parameters of the work rate-time relationship [Abstract]. Med Sci Sports Exerc 1995; 27:S15.
45. Hicks AL, Kent-Braun J, Ditor DS. Sex differences in human skeletal muscle fatigue. Exerc Sports Sci Rev 2001; 29(3):109–112.
46. Cox G, Mujika I, Tumilty D, Burke L. Acute creatine supplementation and performance during a field test simulating match play in elite female soccer players. Int J Sport Nutr Exerc Metab 2002; 12:33–46.
47. Dawson B, Vladich T, Blanksby BA. Effects of 4 weeks of creatine supplementation in junior swimmers on freestyle spring and swim bench performance. J Strength Cond Res 2002; 16(4):485–490.
48. Grindstaff PD, Kreider R, Bishop R, et al. Effects of creatine supplementation on repetitive sprint performance and body composition in competitive swimmers. Int J Sport Nutr 1997; 7:330–346.
49. Peyrebrune MC, Nevill ME, Donaldson FJ, Cosford DJ. The effects of oral creatine supplementation on performance in single and repeated sprint swimming. J Sport Sci 1998; 16:271–279.
50. Hoffman JR, Stout JR, Falvo MJ, Kang J, Ratamass NA. Effect of low-dose, short-duration creatine supplementation on anaerobic exercise performance. J Strength Cond Res 2005; 19(2):260–264.
51. Aoki MS, Gomes RV, Raso V. Creatine supplementation attenuates the adverse effect of endurance exercise on subsequent resistance exercise performance [Abstract]. Med Sci Sports Exerc 2004; 36(5):S334.
52. Nelson AG, Day R, Glickman-Weiss EL, Hegsted M, Kokkenen J, Sampson B. Creatine supplementation alters the response to a graded cycle ergometer test. Eur J Appl Physiol 2000; 83:89–94.
53. Labotz M, Smith BW. Creatine supplement use in an NCAA Division I athletic program. Clin J Sport Med 1999; 9:167–169.
54. Greenwood M, Farris J, Kreider R, Greenwood L, Byars A. Creatine supplementation patterns and perceived effects in select Division I collegiate athletes. Clin J Sport Med 2000; 10:191–194.
55. Bemben MG, Lamont HS. Creatine supplementation and exercise performance: Recent findings. Sports Med 2005; 35(2):107–125.

56. Tarnopolsky MA. Gender differences in metabolism: Nutrition and supplements. J Sci Med Sport 2000; 3(3):287–298.
57. Giacomoni M, Bernard T, Gavarry O, Altare S, Falgairette G. Influence of the menstrual cycle phase and menstrual symptoms on maximal anaerobic performance. Med Sci Sports Exerc 2000; 32(2):486–492.
58. Doolittle TL, Engebretsen J. Performance variations during the menstrual cycle. J Sports Med 1972; 12:54–58.
59. DeBruyn-Prevost P, Masset C, Sturbois X. Physiological response from 18–25 years women to aerobic and anaerobic physical fitness tests at different periods during the menstrual cycle. J Sports Med 1984; 24:144–148.
60. Lebrun CM. Effect of different phases of the menstrual cycle and oral contraceptives on athletic performance. Sports Med 1993; 16:400–430.
61. Oosthuyse T, Bosch AN, Jackson S. Cycling time trial performance during different phases of the menstrual cycle. EJAP 2005; 94(3):268–276.
62. Wehnert-Roti ML, Clarkson PM. Effects of creatine supplementation in women during the luteal and follicular phases of the menstrual cycle [Abstract]. Med Sci Sports Exerc 2000; 32(5):S355.
63. Pearson D. Feed first – supplement second. Strength Cond J 1999; 21(6):67.

5 Creatine Consumption in Health

Jacques R. Poortmans, PhD
and Marc Francaux, PhD

1. INTRODUCTION

Concerns about the deleterious consequences of oral creatine (Cr) supplementation were initiated in Spring 1998. Two British nephrologists published a paper in "The Lancet" suggesting that there is "strong circumstantial evidence that Cr was responsible for the deterioration in renal function" *(1)* (details given in Section 3.3.2). Three days after this publication a French sport newspaper "L'Equipe" (28th April 1998) stressed that Cr is dangerous for the kidneys, in any condition. This news was handed over to several European newspapers. Cr became the champion's viagra with eventual death! Indeed, Pritchard and Kalra commented on the case of three American college wrestlers who died *(1)*. This later turned out to be false and the Food and Drug Administration (FDA) ruled out Cr supplementation as a primary cause of the deaths of these young athletes *(2)*.

A website of the "Food and Drug Administration" (http://vm.cfsan.fda.gov/~dms/aems.html) gives regularly reported complaints from voluntary consumers or healthcare professionals. The search of 20th October 1998 revealed 32 matches for Cr. They specify: dyspnea, fatigue, grand mal seizure, intracerebral haemorrhage, vomiting, diarrhoea, nervousness and anxiety, polymyositis, myopathy, rabdomyolysis, severe stomach cramps, deep vein thromboses, atrial fibrillation, cardiac arrhythmia, chest pain, and death! Potential adverse effects of oral Cr supplementation was also critically reviewed by Juhn and Tarnopolsky who concluded in 1998 that future studies should include large randomized controlled trials evaluating the short- and long-term effects on the renal

From: *Essentials of Creatine in Sports and Health*
Edited by: J. R. Stout, J. Antonio and D. Kalman © Humana Press Inc., Totowa, NJ

and hepatic systems, as well as the many other organ systems in which Cr plays a metabolic role *(3)*. Furthermore, the FDA asks the reader to keep in mind "there is no certainty that a reported adverse event can be attributed to a particular product or ingredient. The available information may not be complete enough to make this determination." In other words there is no scientific evidence to correlate oral Cr monohydrate supplements with any of these reported adverse effects. Nevertheless, the FDA, the "Association of Professional Team Physicians," and the "American College of Sports Medicine" *(4)* concluded that, although short-term studies look positive for oral Cr supplementation, much more long-term research needs to be done before one can issue a verdict on its health status. In its consensus statement, the "American College of Sports Medicine" said that "the fact that Cr is a naturally occurring compound does not make supplementation safe, as numerous compounds are good, even essential in moderation, but detrimental in excess. The lack of adverse effects does not equal safety, since unending research must be performed to eliminate the possibility of all potential complications" *(4)*. In doubt anyone should be left to its free choice and interpretation. In 2004, a Scientific Panel of the "European Food Safety Agency" concluded that "the safety and bioavailability of Cr, Cr monohydrate, in food for particular nutritional uses, is not a matter of concern provided there is adequate control of purity of this source of Cr" *(5)*. However, that same year, a report of the "Agence Française de Sécurité Sanitaire et Alimentaire (AFSSA) claimed that "one should not encourage publicity of Cr in order to protect sport participants to any potential pathological consequences" *(6)*. Eventually, a world expert on Cr metabolism, M. Wyss, came to the conclusion that there are still many open questions related to Cr metabolism, which are worth being analyzed in detail *(7)*.

As will be observed later in this chapter, there are still several reports or reviews pointing out the uncertainties related to health hazard of Cr utilization in sport events or training. Moreover, commercials are adding ahead many positive allegations without any definite or formal evidences.

2. MODIFICATIONS INDUCED BY CR SUPPLEMENTATION WITHIN HEALTHY CONDITIONS

2.1. Total Body Mass

One of the purported effects of oral Cr supplementation regularly mentioned by the consumers is the increase in total body mass (BM) with special attention to muscle mass. On scientific grounds, the situation is less clear. Indeed the average increase in BM reported in the literature amounts to 1–2 kg or 1–2.3% of total BM (Table 1). However, one has to explore the effects of short-term (<10 d), and medium-term (>10 d) on BM. In fact, the short-term supplementations are based on a mean 20 g/d whereas most of the medium-term supplements are consumed as 2–10 g/d. Table 1 emphasizes that the difference between short-term and medium long-term is not so important. The reported literature indicates 75% and 71%, respectively. Thus, despite the reduced daily charge of Cr supplements, most of the reports stressed the increase in BM. Nevertheless about 30% of published papers do not report any change in BM either after short-term or medium medium-term oral Cr supplementation.

Tentative explanations may be put forward to explain this observation on BM: the characteristics of the subjects and the supplementation protocol. Sedentary people, physically active individuals, and recreational athletes seem to respond equally to oral Cr supplementation. Thus, individuals need not be physically active to increase BM to oral Cr supplementation despite different daily energy intake. It cannot be concluded from specific studies on female vs male subjects. Moreover, none of these few studies controlled the menstrual cycle of the female subjects. Presently, it is difficult to postulate a sex dependence as there is no scientific evidence to support those differences. Eventually, not all studies have been made using a control group (no training session) in addition to a placebo and Cr supplementation groups.

Also it will be investigated if the increase in total BM could be attributed to the effect of training itself (*see* Table 1). Apparently, no statistical differences were found in BM in inactive control subjects followed during 9 wk as compared with active people involved in resistance training for the same period of time *(8)*. Thus, even during endurance training, the increase in BM was clearly related to Cr supplementation.

2.2. Free-Fat Mass

Of course, the increase in total BM should be in favor of a more specific characteristic for the athlete; namely free-fat mass (FFM). A dozen studies did investigate the modifications of FFM after Cr supplementation (Table 2). As expected, 11 out of 12 studies reported a higher increase of FFM, as compared with total BM. The increase may even reach 6% after sustained Cr load.

The increase in FFM could be attributed to carbohydrate supplements added to Cr ingestion in several studies. Indeed, water binding to muscle and liver glycogen occurs after glucose loading. Green et al. *(9)* and Robinson *(10)* observed that whole body- and muscle-Cr retention was increased when large amounts of simple carbohydrates were ingested in

Table 1
Cr Supplements and BM Changes

Gender	Population	Dose (g/d)	Duration (d)	Effect on BM(% change)	References
Short-term (<10 d)					
M	Active	24	6 d	+1.5	*140*
M	Trained	20	6 d	+1.2	*141*
M	Active	20	5 d	+1	*142*
M	Active	20	5 d	+1.3	*143*
M	Active	20	6 d	+1.4	*144*
M	Active	20	5 d	+1	*145*
M	Sedentary	20	5 d	+1.1	*9*
M, F	Swimmers	20	5 d	+1.1	*146*
F	Active	0.5/kg body wt	6 d	Stable	*147*
M	Weight lifters	20	5 d	+2.3	*148*
M, F	Cyclists	20	5 d	Stable	*149*
M, F	Swimmers	21	9 d	Stable	*150*
F	Active	25	7 d	Stable	*151*
M, F	Active	18.8	5 d	Stable	*152*
M	Football	21	5 d	Stable	*153*
F	Runners	20	5 d	Stable	*154*
F	Active	20	4 d	Stable	*51*
M	Runners	0.35/kg body wt	3 d	+2	*14*
M, F	Active	20	5 d	Stable	*90*
M	Active	11.4	5 d	+2.3	*91*
M	Karatekas	20	5 d	+1.3	*155*
M	Active	30	5 d	+1.3	*156*
M	Active	20	5 d	+1.4	*10*
M	Trained	25	7 d	+2.1	*157*
M	Westlers	20	5 d	+1.3	*158*
M	Active	20	5 d	+1	*159*
M	Active	20	5 d	+1.8	*58*
M	Active	5	5 d	+2.3	*160*
M	Weight lifters	20	5 d	+0.9	*160*
M	Active	20	5 d	+1.9	*161*

(Continued)

Table 1 *(Continued)*

Gender	*Population*	*Dose (g/d)*	*Duration (d)*	*Effect on BM(% change)*	*References*
F	–	–	–	+0.6	–
M	Soccer	20	6 d	+0.8	*83*
M	Senior	20	5 d	+0.5	*17*
M	Active	20	5 d	Stable	*162*
M	Active	21	7 d	+1.4	*163*
M	Rowers	20	6 d	+1.9	*164*
M	Football	20	5 d	+3.4	*15*
M	Active	20	6 d	Stable	*165*
M	Triathletes	20	5 d	+1.1	*166*
M	Active	20	4 d	+1.5	*87*
M	Active	20	5 d	+0.8	*167*
M	Active	20	7 d	+0.2	*168*
M, F	Runners	0.35/kg body wt	3 d	+1.1	*169*
M	Active	20	5 d	+1.4	*170*
M	Weight lifters	20	4 d	+2.2	*81*
M	Handball	20	5 d	+3.4	*82*
M	Active	20	5 d	+1.2	*94*
M	Active	20	5 d	+1.8	*171*
M	Active	0.25/kg body wt	7 d	+2.7	*16*
M	Swimmers	20	8 d	+2	*172*
M	Active	20	5 d	+1.6	*127*
M, F	Active	20	7 d	+2.4	*173*
M	Active	21	5 d	Stable	*174*
M	Swimmers	20	5 d	Stable	*175*
Medium and long-term (>10 d)					
M	Weight lifters	20	14 d	+1.9	*176*
F	Swimmers	2	42 d	Stable	*177*
M	Football	3	14 d	+0.8	*178*
M, F	Active	0.3/kg body wt	42 d	+2	*179*
F	Active	10.5	51 d	Stable	*153*
F	Active	5	60 d	Stable	*51*
M	Active	3	47 d	+1.8	*180*

(Continued)

Table 1 *(Continued)*

Gender	*Population*	*Dose (g/d)*	*Duration (d)*	*Effect on BM(% change)*	*References*
M, F	Active	3	47 d	Stable	*90*
M	Football	16	28 d	+1	*50*
M	Active	3	63 d	+2.9	*8*
M, F	Swimmers	10	14 d	Stable	*181*
M	Football	8	35 d	+1.4	*182*
M	Active	7	77 d	+6.3	*157*
M	Senior	20	30 d	+0.6	*183*
M	Active	21	14 d	Stable	*184*
M	Active	0.1/kg body wt	21 d	+2.4	*185*
M	Active	0.03/kg body wt	365 d	+10.6	*64*
M	Active	0.1/kg body wt	42 d	+0.9	*92*
M	Senior	0.3/kg body wt	84 d	+3.4	*186*
M	Weight lifters	10	56 d	+5.4	*72*
M	Active	6	84 d	+5.8	*43*
M, F	Active	5	14 d	–	*187*
M	Senior	5	180 d	Stable	*188*
M	Senior	5	98 d	+1.2	*187*
F	–	–	–	+1.7	–
M	Active	5	30 d	+1.3	*189*
M	Active	0.2/kg body wt	42 d	+4	*190*
F	–	–	–	Stable	–
M	Active	0.3/kg body wt	28 d	+2.9	*191*

conjunction with Cr. However, the conclusions of the two studies are not similar. Green reported that Cr loading (20 g daily during 5 d) increased total BM by 0.6 kg but when 370 g of glucose was added over the course of the day the subjects gained an additional 1.5 kg *(9)*. On the contrary, Robinson did not observe significant total BM change after 5 d of 277 g glucose supplements but did report a 1.4% increase (1 kg) postsupplementation with Cr + carbohydrate *(10)*.

Table 2
Cr Supplements and FFM in Humans

Gender (Nb)	*Population*	*Dose (g/d)*	*Duration (d)*	*Effect on FFM (% increase)*	*References*
F (20)	Active	5	70	+5.7	*51*
M (10)	Active	5	5	+3.9	*160*
M (10)	Active	20	5	+2.2	*160*
M (24)	Active	0.1/kg body wt	21	+2.8	*185*
M (7)	Active	2	42	Stable	*161*
F (8)	–	20	5	Stable	–
M (9)	Active	20	5	+3.7	*15*
M (16)	Senior	0.3/kg body wt	84	+6.1	*186*
M (6)	Active	6	84	+4.4	*43*
M (11)	Active	10	56	+6.4	*72*
M (10)	Active	20	4	+3	*81*
M (23)	Active	5	180	+1.5	*188*
		5	365	+1.9	
M (5)	Active	5	98	+2.5	*192*
F (5)	–	–	–	+5.9	–
M (9)	Active	0.3/kg body wt	28	+3.4	*191*

It has been demonstrated that supraphysiological circulating concentrations of insulin, as induced by glucose ingestion, are required to augment muscle-Cr accumulation in humans *(11)*. Consistent with this observation, Willott et al. *(12)* reported that the highly insulin-sensitive rat soleus muscle had higher rates of Cr uptake than the relatively less insulin-sensitive extensor digitorum longus muscle. However the same authors pointed out that insulin had no direct effect on ^{14}C-labeled Cr uptake rates at concentrations in which effects are seen on glucose uptake, glycogen synthesis, and glycolysis. Moreover, using a competitive inhibitor (β-guanidoproprionic acid) and low-extracellular Na^+ concentrations to study the rate of ^{14}C-labeled Cr uptake in isolated rat muscle, Willott et al. suggested that the rate of Cr uptake is strongly dependent on the extracellular Na^+ concentration, insulin playing only a minor role in the regulation in Cr transfer *(12)*.

2.3. Skeletal Muscle Mass

The increase in FFM could be attributed more specifically to muscle mass volume changes. There are a few recent reports that investigate muscle volume changes using either anthropometric dual energy X-ray absorptiometer (DEXA), electrical bioimpedance, or magnetic resonance imaging techniques after Cr supplementation (Table 3). Local muscle groups were measured after resistance training under Cr load. In most cases, the direct effect on muscle volume was observed with a mean increase of 12% when supplementation and training were maintained for several weeks. Thus, one may conclude that oral Cr supplementation has a direct effect on muscle mass volume. However, this increase may be because of water retention in the muscle and/or to real accretion of muscle protein.

2.3.1. Water Distribution in Muscle Tissue

The increase in muscle mass by Cr may be because of water retention in the intracellular compartment or to an increase in dry mass. Hultman et al. *(13)* suggested that the increase in BM during acute Cr feeding is likely to be attributable to body water retention as they observed a 0.6-L decline in urinary volume after ingestion of 20 g of Cr for 6 d.

Multifrequency bioimpedance technique can distinguish between, and assess changes in the body fluid compartments of human subjects (Table 4). Ziegenfuss *(14)* reported 6.6% in tight skeletal muscle volume and a 2–3% increase in total body and intracellular water volume in aerobic and cross-trained men after 3 d oral Cr feeding. In 1999, the incorporation of intra- and extracellular water (by multifrequence bioimpedance) in subjects under Cr supplementation over a period of 9 wk was measured *(8)*. The observed increase in BM (2 kg) could be attributed partially (55%) to an increase in the body water content and more specifically to an increase in the volume of the intracellular compartment (+4.9%). A few more studies repeated these short-term Cr loadings *(15–18)*. All studies but one reported a stable proportion of extracellular water and a 3–8.9% increase of intracellular water. The mechanisms by which Cr supplementation increases intracellular water remains unclear. It is known that Cr transfer into muscle sarcoplasm is governed by a Na^+-dependent, saturable transporter included in the plasma membrane *(19)*. It remains also possible that an osmotic draw of fluid into the intracellular compartment would explain the increase in muscle mass volume. But it may also be suggested that the gain in muscle BM should not be attributed only to water retention, but probably to dry matter growth accompanied with a normal water volume.

Table 3
Cr Supplements and Muscle Volume Changes

Gender (Nb)	*Population*	*Dose (g/d)*	*Duration (d)*	*Exercise (type)*	*Muscle*	*Technique used*	*Effect on muscle (% increase)*	*References*
M (20)	Active	20	5	Resistance training	Thigh	Anthropometry	+6	*185*
M (6)	Untrained	6	84	Resistance training	Thigh	Anthropometry	+18.1	*43*
M (12)	Active	0.25/kg body	7	Power training	Total	DEXA	+1.3	*16*
M (6)	Active	0.2/kg body wt	42	Resistance training	Elbow	DEXA	+12.8	*190*
F (5), M (12)	Active	20	5	Sprint cycling	Thigh	MRI	Stable	*127*
M (12)	Swimmers	20	8	Sprint training	Total	BIA	Stable	*172*
F (6), M (9)	Active	0.3/kg body wt	28	Power training	Total	DEXA	+5.1	*191*

DEXA, dual energy X-ray absorptiometer; MRI, magnetic resonance imaging; BIA, electrical bioimpedance.

Table 4
Cr Supplements and Body Water Distribution Modification

Dose (g/d)	*Duration (d)*	*Total body water (% change)*	*EC water (% change)*	*IC water (% change)*	*References*
0.35/kg body wt	3	Stable	Stable	+3	*18*
3	63	Stable	Stable	+4.9	*8*
20	5	Stable	Stable	Stable	*17*
0.3/kg body wt	7	+0.8	–	–	*88*
20	5	+5.3	Stable	+8.9	*15*
0.25/kg body wt	7	+4.5	+4	+4.9	*16*

EC, extracellular; IC, intracellular.

2.3.2. Protein Muscle Mass

Most, if not all, commercial allegations argued on the basic consequences of sustained creatine supplementation: the increase in skeletal muscle protein mass. Several scientific publications also emphasize the higher content of muscle proteins by indirect implications of the BM or FFM increases. Do we have enough experimental arguments to support this important allegation?

One has to avoid inaccurate mismatch of publications mixing animal studies with human experiments, embryonic and growing models with stable adult situations. Thus, experimental information shall be separated and conclusions refrained to specific models.

2.3.2.1. Animal Studies. The hypothesis of a dry mass increase under Cr supplementation has been already introduced in 1972 by Ingwall et al. *(20)* from in vitro experiments on mononucleated cells and from breast muscle from 12-d chick embryos. Their experiment demonstrated that skeletal muscle cells synthesize myosin heavy-chain faster when supplied with Cr in vitro. The response was apparent within 4 h after addition of Cr to the culture medium and was concentration-dependent over the range of 10–100 μmol (1.3–13 mg/L). Two years later Ingwall et al. showed that Cr stimulated selectively in cultures of differentiating skeletal muscle the rate of synthesis, and not the rate of degradation, of the two contractile proteins, actin and myosin heavy chain *(21)*. The same group extended their previous research on isolated hearts from 17 to 21-d fetal mice maintained in organ culture *(22)*.

Using C2C12 cells, Louis et al. *(23)* were able to confirm that when 5 mmol Cr monohydrate were added to the differentiation medium this supplementation promoted myotube growth whereas guanidopropionic acid depressed it (Francaux, unpublished). This observation postulates that myotube growth could be controlled by the energy status of the cell. Ovine *(24)* and rat *(25)* myogenic satellite cells also appeared to have an effect on growing myofibrils. Both papers suggested that Cr monohydrate supplementation induces differentiation of myogenic satellite cells.

Flisinska-Bojanowska conducted an experiment on rats supplemented with Cr (10 mg/100 g body wt/d) *(26)*. She electrostimulated (50 Hz, 10 min daily for 14 d) the gastrocnemius muscle, and fragments of the white and red portions of the muscle were analyzed for soluble and myofibrillar proteins. The Cr supplemented rats had a 16% increase in myofibrillar proteins, specifically in the white portion of the gastrocnemius when compared with the control group (no Cr, no stimulation). However, it appears that when rats were electrostimulated without Cr supplementation the increase in myofibrillar proteins amounts 50% in the white portion and 37% in the red portion of the muscle. Cr supplementation alone did not change the content of the white muscle but increased the myofibrillar proteins by 18%. So far, these results do not establish a clear-cut on the beneficial effect of a Cr supplements on muscle proteins.

Two further studies do support the conclusion of Ingwall and Flisinska-Bojanowska. Brannon et al. *(27)* investigated the combined effects of Cr supplementation (3.3 mg Cr/g of chow diet) and high intensity run training on the performance capacity and biochemical properties of rodent skeletal muscle. There were no significant changes in either phosphorylcreatine kinase (PCK) activity or myosin heavy-chain isoform distribution following training or supplementation. However, these authors did not give any data on the synthesis of myosin. Fry et al. *(28)* re-examined the effects of Cr supplementation on muscle protein synthesis in tissue culture. They could not support the observation of Ingwall *(20–22)* and Flisinska-Bojanwoska *(26)*. On the contrary, it seems that when adult rats are depleted for Cr through administration of the analog β-guanidopropionic acid, there was a reduction of muscle myofibrillar proteins and atrophy of fiber II *(29,30)*. Adam et al. *(31,32)* investigated the running performance and myosin isomers after β-guanidopropionic acid treatment. This specific drug did not induce any change in running performance but the myosin isomers appear to be reoriented toward the type I phenotype. The interpretation of results obtained in rodents regarding Cr supplementation must be taken with caution. Indeed, although it is generally well accepted that

Cr supplementation increases total muscle-Cr content by about 20% in humans *(13)*, such changes are systematically observed in rodents *(33)*. Moreover, protocols of Cr depletion by β-guanidopropionic acid treatment may not be interpreted as inducing opposite effects of Cr supplementation. Indeed β-guanidopropionic acid can impair ATP level whereas Cr supplementation did not modify muscle-ATP concentration.

Moreover, Murphy et al. *(34)* analyzed the effect of Cr on contractile force and calcium sensitivity in mechanically skinned single fibers from 24- to 28-wk old rat skeletal muscles. They added Cr to the contracting solution in combination with an appropriate volume of water to maintain osmolarity constant and they observed that this solution had beneficial effects on performance of contractile apparatus. This finding suggests that the initial improvement in performance observed with Cr supplementation could be because of a decrease in ionic strength induced by water retention rather than to an energetic effect provoked by higher muscle phophorylcreatine content.

To conclude, there is indirect evidence that Cr supplementation induces muscle protein in vitro and in growing cells and animals.

2.3.2.2. Human Experiments. Already in 1990, Bessman et al. *(35)* suggested that Cr could induce muscle hypertrophy in adult subjects. They founded their hypothesis on the increased uptake of amino acids by the muscle and thereafter an enhanced biosynthesis of myofibrillar proteins. However, they stated that the concentration of Cr *per se* might not be responsible for the stimulation of protein synthesis seen in physiological active muscle. It might well be the increased transport of phosphorylcreatine in the intervening space during contraction that makes more energy available for the ribosomes. Along the same line, it has been shown that cell swelling act as an anabolic signal stimulating protein synthesis and net protein deposition *(36,37)*.

The effect of acute Cr monohydrate supplementation on leucine kinetics and mixed-muscle protein synthesis has been studied more directly in adult human subjects by Parise et al. *(38)*. Young healthy men and women were allocated to Cr (20 g/d for 5 d followed by 5 g/d for 4 d) and tested before and after the supplementation period under rigorous dietary and exercise controls. Intravenous infusion of L-[1-^{13}C]-leucine and mass spectrometry were used to measure mixed-muscle protein fractional synthetic rate and indexes of whole body leucine metabolism. These authors conclude that there was no increase of whole body or mixed-muscle protein synthesis under Cr supplementation. Healthy men were tested before and after oral Cr supplementation (21 g/d during 5 d), myofibrillar protein synthesis in the vastus lateralis and muscle protein breakdown using intravenous infusion of L-[1-^{13}C]-leucine and

L-[$^2H_5$]-phenylalanine, without and with maltodextrin and protein feeding *(39)*. Feeding led to a doubling of myofibrillar protein synthesis and a 40% depression of muscle protein breakdown, but no effect of Cr monohydrate was found on these parameters either in the fed or fast states. Furthermore, the possible stimulatory effect of Cr loading (21 g/d for 5 d) was examined in conjunction with acute resistance exercise on an isokinetic dynamometer (20 × 10 repetitions of leg extension-flexion at 75% of one repetition maximum one leg, before and after Cr intake) *(40)*. Muscle biopsies and arterio-venous differences, under L-[1-^{13}C]-leucine and L-[$^2H_5$]-phenylalanine venous infusion, were used to measure synthetic rates of myofibrillar and sarcoplasmic proteins or muscle protein breakdown. Exercise increased the synthetic rates of myofibrillar and sarcoplasmic proteins by two- to threefold and leg phenylalanine balance became more positive, but Cr loading was without any anabolic effect. Clearly, both exercise in itself and food are much more stronger stimuli for protein synthesis than Cr intake in healthy adult individuals *(41,42)*.

However, a few studies have recently utilized different technical tool of molecular biology to investigate the membrane Cr transporter, muscle-specific gene expression, and some regulatory signals of protein synthesis. Willoughby and Rosene investigated the effect of oral Cr on myosin heavy chain expression in adult male subjects after resistance training *(43)*. Their results on 12 wk Cr supplementation suggested that the expression of myosin heavy chain mRNA are reflected in the observed increase in myofibrillar protein content. Additionally, Deldicque et al. *(44)* investigated the effect of Cr supplementation on insulin-like growth factor (IGF-I and -II) mRNA expression, including the PI3K-Akt/PKB-mTOR signaling pathway, in adult human skeletal muscle. IGF-I and -II mRNA were slightly, but significantly, increased after Cr supplementation (5 d, 21 g/d). IGFs stimulate the PI3K-Akt/PKB-mTOR signaling pathway, which is involved in the regulation of skeletal muscle fiber size and in the stimulation of translation initiation by activating mTOR and two of its downstream effectors, namely p70^{s6k} and 4E-BP1 (eukaryotic initiation factor-4E binding protein-1). The subjects were submitted to a resistance exercise session consisting in a one-leg knee extension and muscle biopsies were taken before and after the exercise test. Although resistance exercise was shown to increase both IGF-I and -II mRNA, Cr did not potentiate this effect. Three hours after stopping the exercise, Cr supplementation did not induce any change in p70^{s6k} or 4E-BP1 expression, as compared with placebo experiment. However, the phosphorylation of the 4E-BP1 factor displayed a slight increase at the 24 h postexercise under the Cr supplementation condition. These conclusions on Cr transporter mRNA in young and elderly healthy humans are also reported by Tarnopolsky et al. *(45)*. However, Willoughby and Rosene reported a positive effect of oral Cr and resistance training on some myogenic regulatory factor (MRF) expression *(46)*. Their study suggested that 12 wk of Cr supplementation, in conjunction with heavy resistance training, increase the mRNA expression of muscle-Cr kinase by way of pretranslational mechanism, likely owing to the concomitant increases in the expression of myogenin and MRF4.

A recent publication by Olsen investigated the influence of Cr monohydrate (6–24 g/d) or protein (20 g/d) supplementation on satellite cell frequency and myonuclei number in healthy adult men during 16 wk of heavy resistance training *(47)*. The results of this study showed that, after 16-wk training, muscle mean fiber area increased by 17% or 8% under Cr or protein supplementation respectively, as compared with 4% in a placebo group. The author concluded that Cr supplementation in combination with strength training amplifies training-induced increase in satellite cell number and myonuclei concentration in adult human skeletal fibers, thereby allowing an enhanced muscle fiber growth in response to strength training.

Eventually, under induced immobilization in adults, Cr supplementation may lead to the expression of muscle myogenic factors as shown by a few papers *(48,49)*. Immobilization of the leg by cast during 2 wk decreased the cross-sectional area by about 10% and maximal knee-extension power by nearly 25%. Oral Cr supplementation stimulated muscle hypertrophy during rehabilitation strength training. This effect appears to be mediated by MRF4 and myogenin expression.

2.4. Muscle Cramp Incidences

Anecdotal reports from athletes have claimed that Cr supplementation may induce muscle cramps (*see* the FAO website). In a recent unpublished study, 12 young healthy males were fed with Cr 3 g/d over a period of 28 d. The subjects were physically active. Five subjects present at least one cramp during sport activities over the supplementation period. Nevertheless, there is no proof to certify that these cramps are directly related to the Cr supplementation. The prevalent hypothesis to explain this potential side was an imbalance in muscle electrolytes. However, Kreider investigated athletes involved in heavy resistance training (5 H/d) for 28 d *(50)*. They were supplemented daily with 15.75 g of Cr monohydrate. There was no evidence of muscular cramping during resistance training sessions or during performance trials. Along the same line, Vandenberghe et al. *(51)* pursued a study on sedentary female subjects who were involved in a 10-wk resistance training with Cr supplementation

(20 g/d for 4 d, then 5 g/d up to 10 wk). No spontaneous side effects were reported during the entire duration of the study. Nevertheless, Juhn et al. *(52)* reported muscle cramping in 25% of 52 baseball and football players who were supplemented with 6–8 g/d during 5 and 3 mo, respectively. A few later studies on 96 young healthy subjects trained during 3 yr *(53)* or on 10 older men *(54)* involved in resistance training did not report cramping incidences owing to Cr supplementation.

The anecdotal reports of muscle cramping might be owing to the intensity of exercise rather than Cr supplementation. Staying well-hydrated could reduce this risk. Moreover, psychological stimulation could foster an individual to exercise over his/her optimal intensity. Meanwhile, further epidemiological studies should be performed to evaluate this potential side effect.

2.5. Gastrointestinal Complaints

Even if rumors sometimes evoke gastrointestinal distress (stomach upset, vomiting, and diarrhoea) among consumers of oral Cr, these assertions are not supported by real evidences. The scientific literature lacks precise information on this matter. Nevertheless, one report from Vandenberghe et al. *(51)* stated that one-third of their subjects (3/9) had minor gastrointestinal distress during 3 d of Cr (40 g/d) and caffeine (400 mg/d) supplementation. As well, Juhn et al. *(52)* reported diarrhoea in 31% of their baseball and football players who were supplemented with 6–8 g of Cr monohydrate during 5 (baseball) and 3 (football) mo, respectively. They suggested that this side effect may be the result of the unusually high osmotic load imposed on the digestive tract of some subjects. On the contrary, Kreider et al. *(50)* did not observe any disturbances among their subjects. Greenhaff supported this observation with his population ingesting 20 g Cr/d *(55)*. However, he mentioned that some discomfort can occur if Cr is incompletely dissolved before ingestion.

In conclusion, there is no reason to believe that oral Cr supplementation had any detrimental effect on the gastrointestinal tract.

3. EVALUATION OF POTENTIAL DETRIMENTAL EFFECTS OF CR SUPPLEMENTATION

3.1. Liver Dysfunctions

Despite the allegations published in sports newspapers and periodicals there are seldom information on liver metabolism changes induced by oral Cr supplementation. Some publications have reported data on liver function while consuming Cr supplements *(50,56–64)*. These results are summarized in Table 5.

Table 5
Cr Supplements and Liver Enzymes in Human Plasma (Mean ± SD)

Enzymes (IU/L)	*Base line*	*Cr*	*Dose (g/d)*	*Duration (wk)*	*References*
ASP	27.5 ± 2.1	32 ± 2.1	20	5	*58*
	42.8 ± 22.6	35.7 ± 10.7	3	4	*62*
	21.3 ± 1.4	22 ± 1.2	10	12	*56*
	19.6 ± 3.7	19.1 ± 1.6	3	9	*62*
	23.1 ± 8.6	26.9 ± 13	15.75	8	*50*
	25 ± 17	32 ± 13	13.7	216	*64*
	26.4 ± 12.9	29.9 ± 8.6	13.9	150	*60*
	34 ± 24	32 ± 22	5–10	76	*50*
	27.2 ± 7.2	29.7 ± 9.2	5	300	*193*
ALT	12.2 ± 1.2	12.7 ± 1.1	20	1	*58*
	27 ± 8	30 ± 7	20	9	*63*
	29.3 ± 15.4	27 ± 8.8	3	4	*62*
	65.8 ± 3.9	70.7 ± 3.7	10	12	*56*
	21.6 ± 2.7	18.7 ± 2.9	3	9	*62*
	24.1 ± 4.7	28.1 ± 9.8	15.75	8	*50*
	24 ± 13	29 ± 15	13.7	216	*64*
	14.5 ± 5.2	14 ± 6.3	13.9	150	*60*
	27 ± 11	27 ± 14	5–10	76	*50*
	21.9 ± 6.1	21 ± 8.9	5	300	*193*
GGT	17.4 ± 2.4	16.4 ± 2.4	20	5	*58*
	27 ± 7	27 ± 9	20	9	*63*
	19.5 ± 5	19.8 ± 7	3	4	*62*
	24.7 ± 2.8	20.1 ± 1.8	10	12	*56*
	24.7 ± 13.2	25.2 ± 15.8	15.75	8	*50*
	27 ± 14	21 ± 9	13.9	150	*60*
	31.4 ± 8.9	28.4 ± 6.7	5	14	*192*
	16.5 ± 6.2	15.3 ± 5.3	5	300	*193*
ALP	72 ± 9	72 ± 9	20	9	*63*
	81.3 ± 18.3	79 ± 16.7	3	4	*62*
	74 ± 13	81 ± 19	13.7	216	*64*
	65.2 ± 15.3	65 ± 17	13.9	150	*60*
	91 ± 29	93 ± 17	5–10	76	*50*
	237 ± 57	209 ± 37	5	300	*193*

ASP, aspartate-oxaloacetate aminotransferase; ALT, alanine-pyruvate aminotransferase; GGT, γ-glutamyl aminotransferase; ALP, alkaline phosphatase.

Indeed, the same group investigated serum enzymes levels, which are of interest for liver origin *(56)*. No changes in enzyme levels were observed during the 8 wk supplementation. Additional information was obtained after oral Cr supplementation in trained subjects (Table 5). No statistical differences were observed throughout the study as far as alkaline phosphatase, aspartate transaminase, alanine transaminase, and γ-glutamyl transpeptidase are concerned. Thus, there is no reason to believe that oral Cr supplementation would induce changes in liver function in human healthy subjects.

However, a report by Duarte et al. *(65)* reported that mice supplemented with oral Cr (0.3 g/body wt) for 6 d had a liver protein content increased by 23%. Among the measured liver enzymes, the authors mentioned that the aspartate transaminase decreased and that the alanine transaminase tended to increase. Moreover, Keys et al. *(66)* reported that mice assigned to 0.05 g Cr monohydrate/kg body wt supplementation for 8 wk underwent chronic hepatitis. Thus, at least in mice, there might be some concern regarding the potential for Cr toxicity. Therefore, Tarnopolsky et al. *(67)* initiated a study to characterize pathological changes of intermediate- and long-term Cr monohydrate supplementation in mice and in rats. They supplemented the animals with 2% (wt/wt) Cr during 1 yr. Histological assessment (20 organs/tissues) was performed on healthy and transgenic (SOD1) mice and in normal rats before and after Cr supplementation. The administration of Cr monohydrate to mice resulted in histological evidence of hepatitis with no evidence of pathology in a variety of other tissues and organs *(67)*. Cr administration to rats did not result in any pathology of all organs/tissues examined. These results clearly show a species- and tissue-specific responses to Cr administration. The authors also insisted that the Cr supplementation were made for one-third to one-half of the life-span of the animals at doses that are those habitually consumed by humans *(67)*.

In addition, several studies on humans *(57,68–72)* did not show any significant increase in plasma urea throughout the duration of Cr supplementation (20 g daily for 5 d or up to 10 g daily for 5 yr) (Table 6). Meanwhile, Earnest pointed out that Cr supplementation did increase by 17% serum urea in females *(57)*. However, the urea levels remain within the range of a normal population. Earnest et al. *(57)* suggested that chronic high dose of Cr supplementation elicited minimal changes in the markers of hepatic function that were evaluated. Looking at their data and knowing that serum urea is not an accurate representative of liver function the authors of this chapter do not share their conclusion.

Table 6
Cr Supplements and Plasma Measurements (Mean ± SEM)

			Urea (mg/L)		Crn (mg/L)		
Subjects	*Doses (g.d)*	*Duration (d, wk, mo, yr)*	*Pre-Cr*	*Post-Cr*	*Pre-Cr*	*Post-Cr*	*References*
Active	10	5 d	158 ± 7	138 ± 10	11.2 ± 0.5	11.3 ± 0.4	*57*
Active	21	5 d	–	–	9.4 ± 1.6	10.2 ± 1.5	*86*
Active	3	63 d	152 ± 1	151 ± 8	9.6 ± 0.5	9.1 ± 0.8	*71*
Active	<10	5 yr	150 ± 6	147 ± 9	8.8 ± 0.6	8 ± 1.5	*70*
Active	9.7	4 yr	120 ± 18	154 ± 34	9 ± 0.3	13 ± 3[a]	*64*
Active	20	7 d	139	164	12.2	12.2	*68*
Active	10	56 d	192 ± 18	219 ± 75	11.1 ± 3	13.4 ± 3.3[a]	*72*
Active	0.3/kg body wt	7 d	–	–	12.3 ± 0.2	14.03 ± 0.4[a]	*88*
Senior	0.3/kg body wt	7 d	–	–	9.8 ± 1.7	10.8 ± 1.9[a]	*54*
Football	13.9	5.6 yr	–	–	–	12 – 19	*60*
Active	5	19 mo	151 ± 37	155 ± 36	12.3 ± 1	11.6 ± 2	*69*
Active	5	1 yr	–	–	11.3 ± 0.2	12.4 ± 0.4[a]	*188*
Active	5	14 wk	–	–	12.6 ± 0.7	4.2 ± 1[a]	*187*
Active	0.3/kg body wt	4 wk	–	–	16.5 ± 0.9	17.9 ± 1.3[a]	*191*
Active	20	8 d	–	–	10.1 ± 1.2	11.8 ± 2.1	*172*
Active	21	14 d	–	–	11 ± 1.4	8.52 ± 1.2	*95*

[a] $p < 0.05$ (post- vs pre-Cr).

Table 7
Effect of Oral Cr Supplementation on Plasma PCK Activity (Mean ± SD)

		Activity (IU/L)		
Dosage (g/d)	*Duration (d)*	*Baseline*	*Cr*	*References*
10	84	89.3 ± 10.2	97.8 ± 7.8	*56*
10	56	118.8 ± 18.6	181.3 ± 23.6[a]	*56*
20	5	–	No change	*61*
20	5	–	No change	*73*
15.75	28	239 ± 106	609 ± 366[a]	*50*
20	5	–	No change	*74*
5	98	53.3 ± 23.6	107.4 ± 76.4[a]	*187*
20	5	48.3 ± 27.3	32.2 ± 16.4	*75*
5	300	398 ± 253	405 ± 255	*193*

[a]$p < 0.05$ (Cr vs baseline).

3.2. Muscle Markers

PCK is commonly used in clinical pathology as a marker of muscle enzyme efflux and thus muscle dysfunction. Contradictory results were recorded under Cr supplementation (Table 7). Five publications did not report changes in total plasma PCK after 5–84 d of 5–20 g Cr/d supplementation *(56,61,73–75)*. Some studies even followed the increase of plasma PCK under either maxilla isometric contractions or endurance 30-km race with control groups without Cr loading. Apparently, even with exercise-induced muscle damage and muscle soreness, there was no modification of indirect muscle markers under Cr supplementation. Moreover, one study investigated also on two other plasma markers of inflammatory muscle markers (prostaglandin E2 and tumor necrosis factor-α). The results indicated that Cr supplementation reduced cell damage and inflammation after the exhaustive 30-km running.

On the contrary, Kreider et al. *(50)* reported a mild elevation in PCK after 28 d of 15.75 g/d. Nevertheless, it is difficult to get a clear situation about these mild changes in PCK levels because the athletes were practicing heavy training, which might induce muscle enzymes efflux *per se*. Moreover, the plasma PCK exists in different isoforms, (M for muscle, B for brain) namely MM for skeletal muscle, MB for heart, and BB for brain. Today there is no specific report available on the enzyme isoforms that are released from the tissues. Most probably, the slight increase observed in a few reports could be attributed to the skeletal muscle isoform but precise information is needed to confirm this hypothesis.

3.3. Kidney Impairments

Already in 1926, Chanutin investigated the fate of Cr when administered to two subjects during 29 and 44 d with a daily intake of 10–20 g *(76)*. The absorption of Cr appeared to be complete. He found an increased creatinine (Crn) excretion as well as significant positive nitrogen balance. Unfortunately, the excretion was measured for only 1 or 2 d after stopping the administration of Cr. Thus, the carry-over effects on Crn excretion were not determined. Two years later, Rose et al. *(77)* reported that after 49 d of feeding 1 g/d, one man and one woman had a 22–25% increase in the Crn excretion. Hyde *(78)* extended the study of Rose et al. looking at 14 subjects of varying age who were fed 1 g Cr daily for 4–10 wk. Eight of their subjects had increased Crn excretion whereas six individuals did not increase Crn excretion when fed with Cr. Subsequently, Crim et al. *(79)* fed healthy young men with Cr 0.23 g daily for 9 d and 10 g daily for 10 d, consecutively. The subjects were trained on a treadmill (5 d/wk) during the 9-d low-Cr feeding. Crn excretion increased by 10–30% during Cr feeding. There was no significant increase in fecal nitrogen during the Cr-feeding period. Moreover, using oral [^{15}N] Cr feeding in humans, Hoberman et al. *(80)* observed that urine is the only major excretatory route of Cr and Crn. Additionally, sweat loss of Cr collected after exercise was insignificant *(79)*.

More recently, there has been several publications on plasma levels of urea and Crn under Cr supplementation (Table 6). None of the seven reports on plasma urea did observe any modification of urea handling by the kidney. On the contrary, 44% of the publications on plasma Crn level did show a mean 15% increase after Cr supplementation. However, there does not seem to be any relationship between the daily load, the duration of supplementation, and the observed slight plasma increase, which mostly remains with the normal range of a healthy population.

Several publications investigated the modifications of the excretion of urea and Crn after Cr supplementation. Some showed an increase in Crn excretion when individuals were fed with Cr *(13,51,81–85)*. However, several authors did not observe any statistical changes in Crn excretion after short-term *(68,86–89)*, medium-term *(71,72)*, or long-term *(69,70)* oral Cr supplementation in trained individuals (Table 8). The urine output was also measured after Cr supplements in some publications and the results are contradictory, either an increase *(90)*, a stable output *(71,91)*, or a decline in urinary volume *(13)*. Hultman et al. *(13)* suggested that the increase in BM during acute Cr loading is likely to be attributable to body water retention. This explanation does not follow from the other studies.

Urea output was also taken into consideration in a few studies. No modification in 24 h urea was observed after 4–7 d *(51,68,81,86)*,

Table 8

Cr Supplements and Urine Measurements in Humans (Mean ± SEM)

Dose (g/d)	Duration (d, mo, yr)	Subjects	Urea (g/24 h) Pre-Cr	Urea (g/24 h) Post-Cr	Crn (g/24 h) Pre-Cr	Crn (g/24 h) Post-Cr	References
20	6 d	Active	–	–	0.99	1.36[a]	13
0.25/kg body wt	5 d	Active	–	–	1.41 ± 0.53	2.09 ± 0.66[a]	85
21	5 d	Active	18.8 ± 2.3	22.2 ± 1.1	1.19 ± 0.08	153 ± 0.005	86
20	4	Active	27.5 ± 1.1	26.8 ± 1.8	1.86 ± 0.08	2.25 ± 0.14[a]	51
3	58 d	Active	12.6 ± 2.8	10.3 ± 2.1	1.73 ± 0.14	1.90 ± 0.10	71
9	5 d	Active	–	–	1.63 ± 0.35	2.21 ± 0.26[a]	84
3–10	8 mo–5 yr	Active	11.3 ± 0.9	12.3 ± 1.3	1.80 ± 0.07	1.53 ± 0.12	70
20	6 d	Active	–	–	2.15 ± 0.69	2.62 ± 0.60[a]	83
10	56 d	Active	41.4 ± 27.8	43.3 ± 23.1	1.38 ± 0.67	1.85 ± 0.77	72
0.3/kg body wt	7 d	Active	–	–	2.34 ± 0.36	2.72 ± 0.73	88
20	4 d	Active	–	–	1.71 ± 0.09	1.85 ± 0.20	87
20	7 d	Active	47.2	38.2	1.87	1.81	68
20	6 d	Active	–	–	1.41 ± 0.53	2.09 ± 0.66[a]	82
20	4 d	Active	41.6 ± 5.6	45.5 ± 9	1.30 ± 0.20	1.80 ± 0.20[a]	81
5	19 mo	Active	–	–	2.82 ± 1.6	2.67 ± 1.1	69
0.06/kg body wt	7 wk	Active	–	–		No change	16
20	8 d	Active	–	–	1.56 ± 0.39	2.71 ± 1.04	172
21	14 d	Active	–	–	1.86 ± 0.14	2.22 ± 0.14	95
5	20 wk	Active	–	–	1.57 ± 0.10	1.64 ± 0.07	134

[a] $p < 0.05$ (post- vs pre-Cr).

Table 9
Cr Supplement Excreted vs Oral Doses

Dose (g/d)	*Duration (d)*	*Excreted (%)*	*References*
10	10	73	*76*
20	10	67	*93*
20	5	67	*13*
0.25/kg body wt	5	57	*85*
21	5	60	*86*
10	5	44	*91*
9	5	33	*84*
25	5	72	*98*
20	1	67	*97*
0.1/kg	7	46	*92*
20	5	55	*94*
20	5	47	*96*
21	14	77	*95*

9–10 wk *(71,72)*, or 10 mo–5 yr *(69,70)* of oral Cr supplementation (Table 8). Therefore, it seems reasonable to say that the liver does not seem to be overproducing urea production when healthy subjects ingest Cr in excess.

As mentioned earlier, Cr load (from 2 to 20 g/d) seems to be totally absorbed by the intestinal tract. However, skeletal muscle cannot take up all this excess Cr and Cr must be excreted in the urine *(13,71,76,84–86,91–98)*. The excreted Cr represents 40–72% of the original load (Table 9).

An early report from Earnest et al. *(57)* stated that Cr supplementation had minimal changes in the markers of renal function. Unfortunately, they used plasma urea, which does not represent a valuable marker when taken without any urine determination to assess renal function. More troublesome, their data did not show any differences in plasma urea for male subjects and a modest increase (17%) for female consumers.

The first investigations on renal functions in healthy individuals who consumed oral Cr supplementation was published 7–9 yr ago *(70,71,86)*. Renal clearances of Crn, urea, and albumin were compared in three different groups of active subjects who consumed Cr during 5 d, 9 wk, and up to 5 yr as compared with control groups. Statistical differences were not observed between the control groups and the Cr consumers (Table 10). Lately, several other investigations supported the authors' primary results on glomerular filtration rate

Table 10
Cr Supplements and GFR in Humans (Mean ± SEM)

Gender				Duration	Effect on GFR (mL/min)		
(M, F, Nb)	*Age (yr)*	*Population*	*Dose (g/d)*	*(d, mo, yr)*	*Pre-Cr*	*Post-Cr*	*References*
M (5)	25.1	Active	21	5 d	130 ± 16	138 ± 25 (2)	*86*
M (20)	21	Active	3	58 d	130 ± 16	143 ± 11 (2)	*71*
M (8)	24	Active	3–10	8 mo, 5 yr	145 ± 8	143 ± 11 (2)	*70*
F (1), M (10)	28.4	Active	20	5 d	124 ± 9	125 ± 6 (2)	*58*
M (7)	22.4	Active	20	5 d	137 ± 18	129 ± 23 (2)	*161*
F (8), M (10)	20.5	Football	13.5	35 mo	124	126 (1)	*60*
M (12)	19.2	Football	15.75	5 d	162 ± 15	120 ± 18 (3)	*69*
M (25)	19.2	Football	5	9.1 mo	162 ± 15	168 ± 33 (3)	
M (17)	19.2	Football	5	1.6 yr	162 ± 15	177 ± 44 (3)	
M (17)	21	Active	5	23 d	109 ± 2.2	110 ± 2.7 (4)	*197*
M (20)	24.1	Active	21	14 d	127 ± 6	137 ± 10 (2)	*95*

Determination of GFR by:
(1) Crn, colorimetric method (urine, plasma).
(2) Crn, enzymatic assay (urine, plasma).
(3) Crn, HPLC (plasma).
(4) Iohexol plasma clearance.

(GFR). From these experimental protocols it can be stated that GFR and the tubular reabsorption process were not affected by oral Cr supplementation using the usual daily amount (20 g/d for 5 d, less than 10 g/d thereafter). However, the use of Crn clearance to assess GFR in healthy athletes who consume oral Cr monohydrate has been criticized by Kuehl et al. *(99)*. In a recent communication at the annual meeting American College of Sports Medicine these authors investigated athletes consuming oral Cr supplements (10 g/d during 56 d) looking at their GFR using the Crn and the iohexol clearances *(100)*. There was a 0.99 correlation between the two methods and all values tracked the same pattern. The authors are thus confident to use the Crn clearance, which is less invasive and more practical to assess impairment of the filtration process at the renal side. A large survey (100 subjects) made by Richard B. Kreider on regular users of Cr who consumed the supplement during 1 yr reached the same conclusion *(101,102)* (and Kreider 2000, personal communication).

Furthermore, the specific immunochemical techniques did not observe any modification induced by Cr loading on urine albumin excretion rate, which remained within the physiological range for healthy subjects *(70,71,86,95)*.

Microalbuminuria is a well-known predictor of kidney impairment *(103)*. The excretion rate of plasma albumin in urine has been widely used to assess increased glomerular membrane permeability in many pathological conditions *(104–106)*. A subclinical increase in urinary albumin excretion rate is a powerful predictor of the later development of persistent proteinuria and renal failure. The upper level of albumin excretion in a healthy population under resting condition is 20 μg/min. Figure 1 shows the values obtained under different conditions of oral Cr supplementation in healthy subjects. None of the 52 subjects show any increase of albumin excretion when compared with a placebo investigation or a control population *(70,71,86,95)*. Thus, it may be stated that the glomerular membrane permeability is not affected by these different loads of Cr monohydrate supplementation in healthy subjects.

Anecdotally, a recent publication of Groeneveld et al. *(107)* on the long-term (310 d) Cr supplementation (10 g/d) in 57 patients with the neurodegenerative disease amyotrophic lateral sclerosis may also be reported. Long-term Cr supplementation did not lead to an increase of plasma urea levels or to a higher prevalence of microalbuminurea (<20 μg/min).

3.3.1. Animal Studies

Because Cr supplementation raised concern regarding its effect on the kidney, Edmunds et al. *(108)* decided to use an animal model to

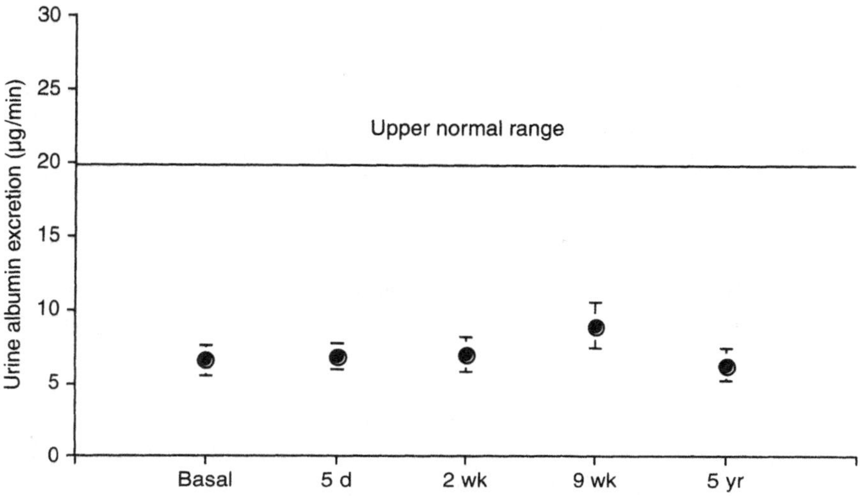

Fig. 1. Urine albumin excretion rate, before and after Cr supplementation *(70,71, 86,95,139)*.

investigate the renal disease progression in Han:SPRD-cy male and female rats. The Han:Sprague-Dawley Renal Disease-cy rat is a well-documented and accepted animal model of inherited renal cystic disease that resembles human autosomal dominant polycystic kidney disease. The authors reasoned that if Cr supplementation affects the kidney in any way, these alterations would be more detected in an animal model that has shorter life-span than humans. Four-week old rats were supplemented daily with 0.4 g Cr/kg body wt during 5 wk. Edmunds et al. *(108)* recorded that Cr supplementation resulted in increased disease progression and worsened renal function in the animal model of kidney disease. They concluded that Cr should be used with particular caution in humans with or at risks for renal disease.

These results on inherited renal cystic disease in rats were not confirmed by another team of nephrologists who investigated normal adult rats submitted to Cr monohydrate (2% wt/wt) during 4 wk *(109)*. Rats were allocated to four experimental groups:

1. Sham-operated, normal diet;
2. Sham-operated, Cr diet;
3. Renal failure (two-thirds nephrectomized), normal diet;
4. Renal failure (two-thirds nephrectomized), Cr diet.

The authors measured serum Crn and urea, 24-h urinary albumin excretion, and GFR. Their study could not demonstrate any deleterious effect of Cr supplementation on kidney function in normal rats or in the animal model with pre-existing moderate renal dysfunction.

Recently, Ferreira et al. *(110)* investigated the effect of Cr supplementation (2 g/d/kg body wt during 10 wk) on renal function in endurance trained rats (treadmill at 12 m/min during 1 h/d). They observed a 40% reduction in resting renal blood flow and GFR with Cr vs control and no modification for the 24-h urinary protein excretion. The treadmill exercise itself had no effect on these kidney parameters but the additional Cr load enhanced them by 15%, without modification on protein excretion. The authors concluded that Cr alone induced an important and significant reduction of both renal plasma flow and GFR. Their results contrast with those of Taes et al. *(109)* who used three different method of GFR in addition to albumin excretion rate. More surprisingly, Ferreira et al. did not find any modification of either renal blood flow or GFR induced by exercise itself. Again, as said by Tarnopolsky *(67)* and Kreider *(59)*, because healthy mice nor rats experienced renal or other tissue pathological changes after long-term Cr supplementation, there is no specific reason to believe that Cr adversely affect renal function or health outcomes.

3.3.2. Human Nephropathies

In 1998, Pritchard and Kalra introduced the first case of kidney damage induced after Cr supplementation *(1)*. The 25-yr-old man mentioned in this study presented a focal segmental glomerulosclerosis, 8 yr ago, with frequently relapsing steroid-response nephrotic syndrome. He had required treatment with cyclosporin, a certified nephrotoxic drug, for the last 5 yr to minimize nephrotic relapses. As a soccer-trained individual this man started loading himself in mid-August 1997 with 5 g Cr monohydrate three times per day for 1 wk and then a maintenance dose of 2 g/d that he had been taking for 7 wk. His GFR dropped by 50%. The GFR evidenced a kidney impairment, which was gradually restored to normal value 1 mo after stopping the Cr supplements.

Another case report of interstitial nephritis was published by Koshy et al. *(111)* in a patient having absorbed 20 g of Cr/d for 4 wk. This previously healthy man presented a 4-d history of nausea, vomiting, and bilateral flank pain. The patient was hospitalized with a serum Crn concentration of 2.3 mg/100 mL (normal upper range limit: 1.5 mg) and a urine protein excretion of 472 mg/d (normal upper range limit: 150 mg). A renal biopsy revealed acute focal interstitial nephritis and acute tubular injury. After stopping the Cr supplements his renal function subsequently became normal. This is an anecdotal case out of thousands of regular Cr consumers. Nevertheless, it emphasizes the recommendation to be tested regularly for urinalysis (*see* Heading 4).

Table 11
Renal Dysfunction and Cr Supplements in Humans

Gender	*Age (yr)*	*Sport*	*Original*	*Dose, time (g/d, wk, mo)*	*Consequences*	*References*
Man	25	Soccer	Nephritic syndrome	15, 1 wk	Focal segmental glomerulo-sclerosis	*1*
Man	20	Active	Healthy	20, 4 wk	Interstitial nephritis	*111*
Man	27	Active	Healthy	?, ?	Acute renal failure	*194*
Man	27	Active	Healthy	?, ?	Acute renal failure	–
Man	28	Active	Healthy	?, ?	Acute renal failure	–
Man	20	Weight lifting	Healthy	2, 7 mo	Focal segmental glomerulo-sclerosis	*195*
Man	28	Baseball	Healthy	?, 30 d	Medullary sponge kidney	*196*

d, day; wk, week; mo, month.

Five more anecdotal cases (abstract reports) were introduced in the recent literature (Table 11). However, being modestly critical, these reports are dubious despite the diagnosis of acute renal failure. Either, the individual doses and duration are not reported for three individuals or the remaining two individuals might have consumed additional other unknown substances (steroids?).

Thus, the absence of controlled values does not allow to seriously conclude that Cr supplementation is an inducer of kidney impairment in healthy subjects.

Eventually, Cr supplementation has been administered in dialyzed patients by Kirschbaum (2000, personal communication) without any side effects on blood chemistry.

3.4. Mutagenicity and Carcinogenicity Risks of Excess Cr Supplementation

3.4.1. Methylamine and Formaldehyde Production

Based on the excellent and extended review by Wyss and Kaddurah-Daouk on Cr and Crn metabolism *(112)*, the French Food Agency (AFSSA) claimed unequivocally that excess consumption of Cr and Crn might induce derived carcinogenic and mutagenic compounds, which could put athletes and consumers of exogenous Cr at risk (evaluation of risks induced by Cr consumer and truth on allegations related to sport performance or increase in muscle mass. http://www.afssa.fr).

Indeed, the excess conversion of Cr to sarcosine may result in cytotoxic agents such as methylamine *(112)*. The latter has been found to be deaminated by semicarbazide-sensitive amine oxidase (SSAO, EC 1.4.3.6) to produce formaldehyde and hydrogen peroxide *(113)* (Fig. 2). Under special conditions, methylamine and formaldehyde are two well-known cytotoxic agents, the presence of which can be revealed by urine analyses *(113–116)*. Formaldehyde has the potential to cross-link proteins and DNA, leading to cytotoxicity and carcinogenic effects in cells *(117,118)*. The toxic aldehyde is related to different pathological conditions such as vascular damage, diabetic complications, and nephropathies.

In 2000, Yu and Deng *(119)* administrated a single dose of Cr (50 mg/kg) to mice, which did not seem to alter the urinary methylamine excretion. However, indirect selective inhibition of SSAO activity dramatically induced a fivefold increase in methylamine excretion. The authors concluded that chronic administration of a large quantity of Cr can increase the production of formaldehyde, which might potentially cause serious unwanted side effects on healthy athletes. This conclusion was amplified by the AFSSA, which led the French government to ban any official buying of Cr.

An old German publication showed that exercise practice (one man after a strenuous ski racing) induced a 2.5-fold increase in the urinary excretion of methylamine *(120)*. The authors argued that all conditions associated with creatinuria (such under supplementation during muscular exertion) implicate an increased excretion of methylamine. Of course, in these old days, scientists were not aware of the eventual potential risks of aldehyde formation in human tissues. However, looking at the original publication it was pointed out that the authors used an aliquot of urine obtained after exertion, without any information on the urine output.

Thus, recently 20 male young healthy subjects who were daily supplemented with 21 g Cr monohydrate during 14 d were investigated. Before and after Cr supplementation 24-h urine was collected and Cr,

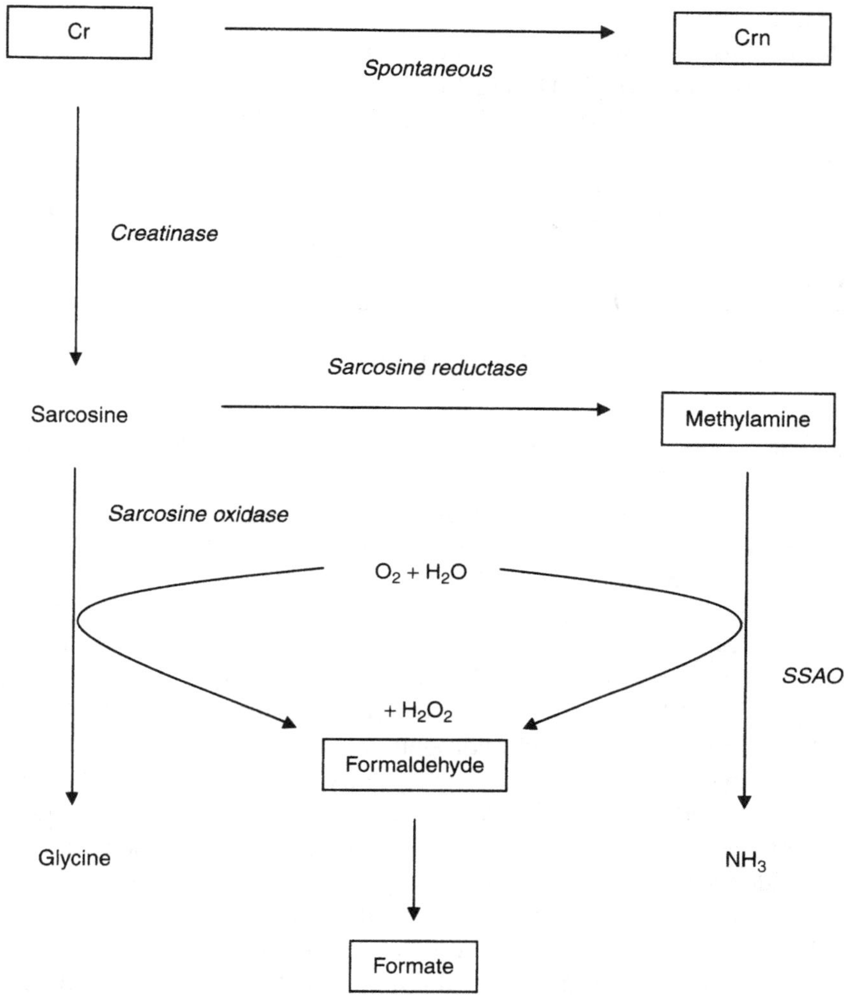

Fig. 2. Schematic pathways of Cr and Crn degradation in the human body. All products and metabolic steps are not shown. Compounds that are framed have been assayed in urine before and after Cr supplementation. SSAO, semicarbazide amine oxidase *(95)*.

Crn, methylamine, formate, and formaldehyde was determined. Table 12 includes the modifications of urine excretion of formaldehyde, formate, and methylamine before and after Cr supplementation. Twenty-four hour urine output of methylamine and formaldehyde increased 9.2- and 4.5-fold, respectively ($p < 0.001$), after Cr feeding with no increase in formate excretion. After Cr feeding, there was no correlation between

Table 12
Mean Values (± SEM) of Urine Contents Before and After Cr Supplementation, 21 g/d During 14 d[a]

	Before Cr	*After Cr*
Methylamine (mg/24 h)	0.69 ± 0.06	6.41 ± 1.45[b]
Formaldehyde (μg/24 h)	64.78 ± 16.28	290.4 ± 66.3[b]
Formate (mg/24 h)	12.46 ± 1.04	14.16 ± 1.84
Albumin (mg/24 h)	9.78 ± 1.93	6.97 ± 1.15

[a] From ref. *95*.
[b] $p < 0.001$ between values before and after Cr supplementation.

plasma Cr and urine methylamine ($r^2 = 0.025$, P = NS) or formaldehyde ($r^2 = 0.017$, P = NS).

The results from the investigation indicate that short-term oral Cr feeding in healthy subjects enhances the mechanisms leading to the conversion of Cr to sarcosine and then to methylamine, the latter one giving rise to formaldehyde. The conversion of formaldehyde to formate should be rather rapid in cells, the latter one representing indirectly the production of the former substrate *(121)*. Using rat and mice models, Yu and Deng *(114,119)* demonstrated that in vivo deamination of methylamine produces formaldehyde and hydrogen peroxide, which are both recognized as cytotoxic substances. Consequently, these authors hypothesized that chronic administration of large quantities of Cr as an ergogenic supplement would increase the production of methylamine and subsequently formaldehyde, both being potentially cytotoxic in renal glomerula *(114,119)*. The results support this hypothesis in humans.

Despite the 9.2-fold increase in methylamine urine excretion induced by Cr ingestion, this level did not reach the normal upper limit values from healthy humans, up to 35 mg/d (mean + 3 SD) *(115)*. After Cr supplementation, urine formate excretion remains below the upper range (14–20 mg/d) reported in healthy subjects *(122–124)*. However, under Cr supplementation, the urine excretion of formaldehyde has been increased 4.5-fold of the basal rate.

Because Cr is transformed to sarcosine by microbial enzymatic reactions *(112)*, it is likely that methylamine is formed in the intestine and is therefore potentially damageable for the integrity of the intestinal epithelium. Methylamine is toxic to human endothelial cells and forms patch-like lesions *(125)* and even kidney damage *(113)*. In mammals, SSAO activity has been found in various tissues associated to vascular system *(126,127)*. Therefore, it is likely that the deamination of methylamine occurs in circulation. It could also be speculated that this flooding of methylamine in blood, together with SSAO, might produce formaldehyde, which favors microangiopathy in the renal glomeruli *(116,127)*.

The subjects consumed a total amount of 280 g of Cr monohydrate over 14 d without any modification of glomerular membrane permeability as assessed by their albumin urine excretion rate (9.78 ± 1.93 mg/24 h before Cr; 6.97 ± 1.15 mg/24 h after Cr). The upper limit of healthy humans is 25 mg/24 h. Albuminuria has long been known to be associated with specific renal abnormality, and is now recognized as an early test for vascular endothelial damage *(128)*. Despite the fact that formaldehyde and methylamine excretion rates were increased respectively to 4.5- and 9.2-fold after Cr supplementation in the subjects, there was no detectible consequence of glomerulonephropathy (Table 11). In this context, it has been shown, at least in rats, that formaldehyde administration in drinking water supplied *ad libitum* during 2 yr can produce specific carcinogenic effects on various organs and tissues *(129)*. This raises the question of the duration of the supplementation. In a previous study, no adverse effect was observed of a long-term (up to 5 yr) Cr supplementation in humans *(70)* (Fig. 1).

Even if systematic deleterious effect could not be observed, it cannot not exclude that a systematic production of low extra doses of cytotoxic agents never induce any nephropathy incidences. Clearly, epidemiological data are required to evaluate potential risks over a larger cohort of individuals. But in terms of results of the present investigation, caution should be applied. Kidney function of the patients and healthy subjects supplemented with Cr on a regular basis should be systematically monitored throughout the ingestion period.

To conclude, the investigation shows that short-term, heavy load oral Cr supplementation stimulates the production of an excess of methylamine and formaldehyde in urine of healthy humans. Even though the production of cytotoxic agents has no apparent effect on the kidney function of volunteers in this study, long-term and epidemiological data are essential to assess whether Cr supplementation is harmless in all healthy individuals under all conditions.

3.4.2. Induction of Carcinogenic and Mutagenic Amino-Imidazo-Azaarene Formation

The review of Wyss and Kaddurah-Daouk *(112)* reported that the processing of foods, in particular frying and broiling of meat, is associated with the generation of mutagenic and carcinogenic substances, namely the amino-imidazo-azaarenes products that shall be named the "heterocyclic amines (HCA)" for simplicity. These substances, which are

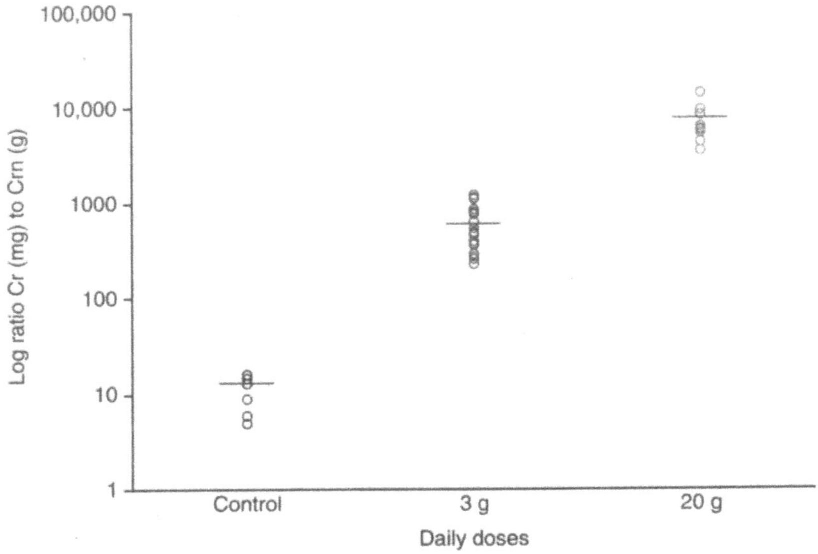

Fig. 3. Urinary excretion of Cr and Crn, expressed as the ration of Cr (mg) to Crn (g) in a urine sample by control individuals and study participants who ingested 3 or 21 g Cr monohydrate per day. The horitontal bars represent the geometric means of each population. A clear separation is observed between nonconsumers and consumers *(62)*.

numerous and classified into five groups (*see* ref. *112* for details), are formed during cooking in the presence of sugar and amino acid, depending on the cooking time and temperature (>250°C) *(130,131)*. Maximal mutagen yield is achieved by mixing Cr or Crn with amino acid and sugar with a molar ratio of 1:1:0.5 *(112,132)*. Crn rather than Cr is likely to be the actual precursor of the HCA mutagens *(112)*. Among the HCA compounds, imidazo-quinoxaline, 8-methyl-imidazo-quinoxaline, 4,8-dimethyl-imidazo-quinoxaline, and imidazo-pyridine are the most important mutagens and together contribute to about 80% of the mutagenicity. HCA are formed at high temperature, during frying or broiling of meat (barbecue effect!) and then at low concentration, so low that it is questionable whether they represent any significant cancer risk. Moreover, at 37°C, HCA formation from Cr or Crn most likely does not occur. Therefore, it would seem very unlikely at present to attribute any cancer risk to oral Cr supplementation *(112)*.

The negative opinion on oral Cr supplementation seem to be purported carcinogenic effect of Cr. Based on current knowledge, the probability that nitrosation products of Cr are formed in the stomach to any significant extent is close to zero *(133)*. A very recent short publication by Derave et al. *(134)* supports this conclusion. These authors investigated in a double-blind, placebo-controlled study the urinary excretion of *N*-nitrososarcosine after 1-wk high dose (20 g/d) and 20-wk low-dose (5 g/d) Cr supplementation in healthy humans. They concluded that Cr ingestion does not increase the urinary excretion of the carcinogen *N*-nitrososarcosine.

The identification of HCA in human urine is not an easy procedure. The analytical methods involved solid-phase extraction and quantification by combined liquid chromatography-tandem mass spectrometry to identify the major HCA in urine *(135–138)*. Nevertheless, one will have to quantify this potential hypothetic risk to definitively exclude unproved allegation still present in nonscientific publications or media.

4. PRACTICAL CONCLUSIONS

The purpose of the present review was to present data and conclusions on the potential side effects of oral Cr supplementation in healthy individuals. Despite papers and editorials published in sports media, there are no real incidents of muscle cramps, gastrointestinal discomfort, or liver impairment after regular load of oral Cr.

There is neither apparent kidney dysfunction when healthy individuals consume oral Cr monohydrate with the usual daily amounts (20 g for 5 d, <10 g afterwards). The few renal incident remain anecdotal. Even if there are no health risks induced by oral Cr supplementation it sounds safe to remain cautious when this substance is administrated on a chronic basis. The excess Cr ingestion is still a burden to be eliminated mostly by the kidney. Regular checkups should be the elementary tactic to follow the potential dysfunction, which could appear with some individuals less prone to compensate any homeostasis imbalance. Blood chemistry for liver enzymes, urea, Crn should be investigated regularly (once a year). The analysis of urinary albumin excretion rate (<20 μg/min) appears to be the most simple, inexpensive, accurate test to assess any early incident of kidney impairment. Should this happen under resting condition, i.e., after 20 h of physical activity, further investigations need to be done by a nephrologist *(139)*.

It is quite easy to determine the Cr consumers when looking at the Cr/Crn ratio from a sample of urine (Fig. 3). This ratio is always less than 40 (mg Cr/g Crn) in nonconsumers and higher than 150 for individuals consuming at least 2 g of Cr/d. Thus, it is believed that healthy subjects are not confronted with health risks when consuming reasonable amounts of oral Cr monohydrate.

Nevertheless, it is advised that Cr supplementation should not be used by a person with pre-existing renal disease or those with a potential risk for renal dysfunction (diabetes, hypertension, and reduced GFR).

Great care should also be taken as far as the purity of exogenous Cr supplements is concerned. Analytical tests must prove their unique nutraceutical composition as safety is not assured in most preparations.

5. CONCLUSION

Doubtful allegations and adverse effects of Cr supplementation have been released from either press media or scientific publications. The present chapter has tried to separate the wheat from the chaff by looking for experimental evidences. One of the purported effects of oral Cr supplementation is to increase muscle mass to improve performance. A review of literature reveals a 1–2.3% increase in total BM after short-term (<10 d) or medium-term (>10 d), respectively. This increase is more focussed to FFM, which might gain a mean 3.3% after several weeks of exogenous Cr load. This increase in FFM is more specifically attributed to skeletal muscle mass changes with a mean 6.2% increase after Cr supplementation. Bioimpedance techniques showed that this increase in muscle mass by Cr is partially because of some water retention in the intracellular compartment (mean + 4.3%) whereas the remaining increase was supposed to be dry mass, such as protein. There is indirect evidence that Cr supplementation induces skeletal muscle protein increase in vitro and in growing cells and animal models. However, experimental investigation in healthy subjects failed to demonstrate any modification of muscle protein synthesis and breakdown under Cr supplementation. Clearly, exercise and amino acid intake are much stronger stimuli for protein synthesis measured over a period of a few hours.

Anecdotal reports from athletes have been released on muscle cramp incidences and gastrointestinal complaints during Cr supplementation, but these few incidences remain rare and not necessarily linked to Cr itself. Despite several allegations from scientific and press media, liver (enzymes, urea) and kidneys (glomerular filtration and albumin excretion rates) are keeping their functionality in healthy subjects supplemented with Cr, even during several months. However, it is advised that Cr supplementation should not be used by a person with pre-existing renal disease or those with a potential risk for renal dysfunction (diabetes, hypertension, and reduced GFR).

Mutagenicity and carcinogenicity potential effects (production of HCA) induced by Cr supplementation have been claimed by a French sanitary agency (AFSSA), which might put consumers at risk. Even if there is a slight increase (within the normal range) of urinary methylamine and formaldehyde excretion after a heavy load of Cr (20 g/d), without any incidence on kidney function, the search for the HCA excretion remains a future task to definitively exclude the unproved AFSSA allegation.

ACKNOWLEDGMENTS

We are indebted to the "Direction Générale des Sports" (French Community of Belgium), to the "Conseil de Prévention et de Lutte contre le Dopage" (Paris, France) for their financial support, and to Flamma SpA (Italy) and Degussa (Germany), which kindly provided the Cr monohydrate.

REFERENCES

1. Pritchard NR, Kalra PA. Renal dysfunction accompanying oral creatine supplements. Lancet 1998; 351:1252–1253.
2. Farquhar WB, Zambraski EJ. Effects of creatine use on the athlete's kidney. Curr Sports Med Reports 2002; 1:103–106.
3. Juhn MS, Tarnopolsky M. Potential side effects of oral creatine supplementation: a critical review. Clin J Sport Med 1998; 8:298–304.
4. American College of Sports Medicine. The physiological and health effects of oral creatine supplementation. Med Sci Sports Exerc 2000; 32:706–717.
5. European Food Safety Acency. Creatine monohydrate for use in foods for particular nutrional uses (Question number EFSA-Q-2003-125). EFSA J 2004; 36:1–6.
6. Agence Française de Sécurité Sanitaire et Alimentaire, Avis relatif à la publicité portant sur des substances de développement musculaire et de mise en forme contenue dans un magazune spécialisé. AFSSA. (Saisines 2003-SA-0385 & 2003-SA-0386) 2004; 1–3. http://www.afssa.fr.
7. Wyss M. Writing about creatine: is it worth the risk? Toxicol Lett 2004; 152:273, 274.
8. Francaux M, Poortmans J. Effects of training and creatine supplement on muscle strength and body mass. Eur J Appl Physiol 1999; 80:165–168.
9. Green A, Hultman E, MacDonald I, Sewell S, Greenhaff P. Carbohydrate ingestion augments skeletal muscle creatine accumulation during creatine supplementation in humans. Am J Physiol 1996; 271:E821–E826.
10. Robinson TM, Sewell DA, Hultman E, Greenhaff PL. Role of submaximal exercise in promoting creatine and glycogen accumulation in human skeletal muscle. J Appl Physiol 1999; 87:598–604.
11. Steenge GR, Lambourne J, Casey A, MacDonald IA, Greenhaff PL. The stimulatory effect of insulin on creatine accumulation in human skeletal muscle. Am J Physiol 1998; 275:E974–E979.
12. Willott CA, Young ME, Leighton B, et al. Creatine uptake in isolated soleus muscle: kinetics and dependence on sodium, but not on insulin. Acta Physiol Scand 1999; 166:99–104.
13. Hultman E, Söderlund K, Timmons J, Cederblad G, Greenhaff P. Muscle creatine loading in men. J Appl Physiol 1996; 81:232–237.

14. Ziegenfuss T, Lemon P, Rogers M, Ross R, Yarasheski K. Acute creatine ingestion: effects on muscle volume, anaerobic power, fluid volumes and protein turnover. Med Sci Sports Exerc 1997; 29:S127.
15. Bemben MG, Bemben DA, Loftiss DD, Knehans AW. Creatine supplementation during resistance training in college football athletes. Med Sci Sport Exerc 2001; 33:1667–1673.
16. Burke DG, Chilibeck PD, Parise G, Candow DG, Mahoney D, Tarnopollsky MA. Effect of creatine and weight training on muscle creatine and performance in vegetarians. Med Sci Sport Exerc 2003; 35:1946–1955.
17. Rawson ES, Clarkson PM. Acute creatine supplementation in older men. Int J Sports Med 2000; 21:71–75.
18. Ziegenfuss TN, Lower LM, Lemon P. Acute fluid volume changes in men during three days of creatine supplementation. J Exerc Physiol 1998; 1:1–7.
19. Guimbal C, Kilimann M. A Na+ dependent creatine transporter in rabbit brain, muscle, heart, and kidney. J Biol Chem 1993; 268:8418–8421.
20. Ingwall J, Morales M, Stockdale F. Creatine and the control of myosin synthesis in differentiating skeletal muscle. Proc Natl Acad Sci 1972; 69:2250–2253.
21. Ingwall J, Weiner C, Morales M, Davus E, Stockdale F. Specificity of creatine in the control of muscle protein synthesis. J Cell Biol 1974; 63:145–151.
22. Ingwall J, Wildenthal K. Role of creatine in the regulation of cardiac synthesis. J Cell Biol 1976; 68:159–163.
23. Louis M, Awede B, Lebacq J, Francaux M. Effect of creatine and guanidino-propionic acid on myotube growth. Med Sci Sport Exerc 2001; 33:S67.
24. Vierck JL, Icenoggle DL, Bucci L, Dodson MV. The effects of ergogenic compounds on myogenic satellite cells. Med Sci Sport Exerc 2003; 35:769–776.
25. Dangott B, Schultz E, Mozdiak PE. Dietary creatine monohydrate supplementation increases satellite cell mitotic activity during compensatory hypertrophy. Int J Sport Nutr 2000; 20:13–16.
26. Flisinska-Bojanowska A. Effects of oral creatine administration on skeletal muscle protein and creatine levels. Biol Sport 1996; 13:39–46.
27. Brannon T, Adams G, Conniff C, Baldwin K. Effects of creatine loading and training on running performance and biochemical properties of rat skeletal muscle. Med Sci Sports Exerc 1997; 29:489–495.
28. Fry D, Morales M. A reexamination of the effects of creatine on muscle protein synthesis in tissue culture. Acta Physiol Scand 1995; 153:207–209.
29. Laskowski M, Chevli R, Titch C. Biochemical and ultrastructural changes in skeletal muscle induced by a creatine antagonist. Metabolism 1981; 30:1080–1085.
30. Van Deursen J, Jap P, Heerschap H, ter Laak H, Ruitenbeek W, Wieringa B. Effects of the creatine analogue β-guanidopropionic acid on skeletal muscles of mice deficient in muscle creatine kinase. Biochim Biophys Acta 1994; 1185:327–335.
31. Adams G, Bodell P, Baldwin K. Running performance and cardiovascular capacity are not impaired in creatine-depleted rats. J Appl Physiol 1995; 79:1002–1007.
32. Adams G, Haddad F, Baldwin K. Interaction of chronic creatine depletion and muscle unloading: effects on postural and locomotor muscles. J Appl Physiol 1994; 77:1198–1205.
33. Louis M, Raymackers JM, Debaix H, Lebacq J, Francaux M. Effect of creatine supplementation on skeletal muscle of mdx mice. Muscle Nerve 2004; 29(5):687–692.

34. Murphy M, Stephenson DG, Lamb GD. Effect of creatine on contractile force and sensitivity in mechanically skinned single fibers from rat skeletal muscle. Am J Physiol 2004; 287:C1589–C1595.
35. Bessman S, Savabi F. The role of the phosphocreatine energy shuttle in exercise and muscle hypertrophy, in Biochemistry of Exercise VII. Taylor A., et al. (eds.), Human Kinetics: Champaign (USA), 1990, pp. 167–178.
36. Berneis K, Ninnis R, Haussinger H, Keller U. Effects of hyper- and hypoosmolality on whole body protein and glucose kinetics in humans. Am J Physiol 1999; 276:E188–E195.
37. Lang F, Busch GL, Ritter M, Volkl H, Waldegger S, Gulbins E, Haussinger H. Functional significance of cell volume regulatory mechanisms. Physiol Rev 1998; 78:247–306.
38. Parise G, Mihic S, MavLennon D, Yarasheski K, Tarnopolsky MA. Effects of acute creatine monohydrate supplementation on leucine kinetics and mixed-muscle protein synthesis. J Appl Physiol 2001; 91:1041–1047.
39. Louis M, Poortmans JR, Francaux M, et al. Creatine supplementation has no effect on human muscle protein turnover at rest in the postabsorptive or fed states. Am J Physiol 2003; 284:E764–E770.
40. Louis M, Poortmans JR, Francaux M, et al. No effect of creatine supplementation on human myofibrillar and sarcoplasmic protein synthesis after resistance exercise. Am J Physiol 2003; 285:E1089–E1094.
41. Paddon-Jones D, Bornsheim E, Wolfe RR. Potential ergogenic effects of arginine and creatine supplementation. J Nutr 2004; 134:2888S–2894S.
42. Rennie MJ, Tipton KD. Protein and amino acid metabolism during and after exercise and the effect of nutrition. Ann Rev Physiol 2000; 20:457–463.
43. Willoughby DS, Rosene J. Effects of oral creatine and resistance training on myosin heavy chain expression. Med Sci Sport Exerc 2001; 33:1674–1681.
44. Deldicque L, Louis M, Theisen D, et al. Increased IGF mRNA in human skeletal muscle after creatine supplementation. Med Sci Sport Exerc 2005; 37:731–736.
45. Tarnopolsky M, Parise G, Fu MH, et al. Acute and moderate-term creatine monohydrate supplementation does not affect creatine transporter mRNA or protein content in either young or elderly humans. Mol Cell Biochem 2003; 244:159–166.
46. Willoughby DS, Rosene J. Effects of oral creatine and resistance training on myogenic regulatory factor expression. Med Sci Sport Exerc 2003; 35:923–929.
47. Olsen S, Aagaard P, Kadi F, et al. Creatine supplementation augments the increase in satelitte cell and myonuclei number in human skeletal muscle induced by strength training. J Physiol 2006; 573:525–534.
48. Hespel P, Op't Eijnde B, Van Leemputte M, et al. Oral creatine supplementation facilitates the rehabilitation of disuse atrophy and alters the expression of muscle myogenic factors in humans. J Physiol 2001; 536:625–633.
49. Op't Eijnde B, Derave W, Wojtaszewski JFP, Richter EA, Hespel P. AMP kinase expression and activity in human skeletal muscle: effects of immobilization, retraining, and creatine supplementation. J Appl Physiol 2005; 98:1228–1233.
50. Kreider R, Ferreira M, Wilso M, Grindstaff P, Plisk S, Reinardy J. Effects of creatine supplementation on body composition, strength, and sprint performance. Med Sci Sports Exerc 1998; 30:73–82.

51. Vandenberghe K, Goris M, Van Hecke P, Van Leemputte M, Vangerven L, Hespel P. Long-term creratine intake is beneficial to muscle performance during resistance training. J Appl Physiol 1997; 83:2055–2063.
52. Juhn MS, O'Kane JW, Vinci DM. Oral creatine supplementation in male collegiate athletes: a survey of dosing habits and side effects. J Am Diet Assoc 1999; 99:593–595.
53. Greenwood M, Kreider RB, Melton C, et al. Creatine supplementation during college footbal training does not increase the incidence of cramping or injury. Mol Cell Biochem 2003; 244:83–88.
54. Gotschalk LA, Volek JS, Staron RS, Denegar CR, Hagerman F, Kraemer WJ. Creatine supplementation improves muscular performance in older men. Med Sci Sport Exerc 2002; 34:537–543.
55. Greenhaff P. Renal dysfunction accompanying oral creatine supplements. Lancet 1998; 352:233.
56. Almada A, Mitchell T, Earnest C. Impact of chronic creatine supplementation on serum enzyme concentration. FASEB J 1996; 10:A791.
57. Earnest C, Almada A, Mitchell T. Influence of chronic creatine supplementation on hepatorenal function. FASEB J 1996; 10:A790.
58. Kamber M, Koster M, Kreis R, Walker G, Boesch C, Hoppeler H. Creatine supplementation – Part I: performance, clinical chemistry, and muscle volume. Med Sci Sport Exerc 1999; 31:1763–1769.
59. Kreider RB. Species-specific responses to creatine supplementation. Am J Physiol 2003; 285:R725–R726.
60. Mayhew DL, Mayhew JL, Ware JS. Effects of long-term creatine supplementation on liver and kidney functions in American college football players. Int J Sport Nutr Ex Metabol 2002; 12:453–460.
61. Mihic S, MacDonald JR, McKenzie S, Tarnopolsky MA. The effect of creatine supplementation on blood presure, plasma creatine kinase, and body composition. FASEB J 1998; 12:A652.
62. Poortmans JR, Francaux M. Adverse effects of creatine supplementation: Fact or fiction? Sports Med 2000; 30:155–170.
63. Robinson TM, Sewell DA, Casey A, Steenge GR, Greenhaff PL. Dietary creatine supplementation does not affect some haematological indices, or indices of muscle damage and hepatic and renal function. Br J Sports Med 2000; 34:284–288.
64. Schilling BK, Stone MH, Utter A, et al. Creatine supplementation and health variables: a retrospective study. Med Sci Sport Exerc 2001; 33:183–188.
65. Duarte JA, Neuparth MJ, Soares JMC, Appell HJ. Oral creatine supplementation and liver metabolism. Int J Sports Med 1999; 20:S50.
66. Keys S, Tyminski M, Davis J, Bacon C, Benglovanni J, Hussin A. The effects of long-term creatine supplementation on liver architecture in mice. Med Sci Sport Exerc 2001; 33:S206.
67. Tarnopolsky M, Bourgeois JM, Snow RJ, et al. Histological assessment of intermediate- and long-term creatine monohydrate supplementation in mice and rats. Am J Physiol 2003; 285:R762–R769.
68. Jowko E, Ostraszewski P, Jank M, Sacharuk J, Zieniewicz J, Nissen S. Creatine and β-hydroxy-β-methylbutyrate (HMB) additively increase lean body mass and muscle strength during weight-training program. Nutrition 2001; 17:558–566.

69. Kreider RB, Melton C, Rasmussen C, et al. Long-term creatine supplementation does not significantly affect clinical markers of health in athletes. Mol Cell Biochem 2003; 244:95–104.
70. Poortmans J, Francaux M. Long-term oral creatine supplementation does not impair renal function in healthy athletes. Med Sci Sports Exerc 1999; 31:1108–1110.
71. Poortmans JR, Francaux M. Renal dysfunction accompanying oral creatine supplements—reply. Lancet 1998; 352:234.
72. Tarnopolsky MA, Parise G, Yardley NJ, Ballantyne CS, Olatinji S, Phillips SM. Creatine-dextrose and protein-dextrose induce similar strength gains during training. Med Sci Sport Exerc 2001; 33:2044–2052.
73. Engelhardt M, Neumann G, Berbalk A, Reuter I. Creatine supplementation in endurance sports. Med Sci Sports Exerc 1998; 30:1123–1129.
74. Rawson ES, Gunn B, Clarkson PM. The effects of creatine supplementation on exercise-induced muscle damage. J Strength Cond Res 2001; 15:178–184.
75. Santos RVT, Bassit RA, Caperuro EC, Costa Rosa LFBP. The effect of creatine supplementation upon inflammatory and muscle soreness markers after a 30km race. Life Sci 2004; 75:1917–1924.
76. Chanutin A. The fate of creatine when asministered to man. J Biol Chem 1926; 67:29–41.
77. Rose WC, Ellis RH, Helming OC. The transformation of creatine into creatinine by the male and female organism. J Biol Chem 1928; 77:171–184.
78. Hyde E. Creatine feeding and creatine-creatinine excretion in males and females of different age groups. J Biol Chem 1942; 143:301–310.
79. Crim MC, Calloway DH, Margen S. Creatine metabolism in men: urinary creatine and creatinine excretions with creatine feeding. J Nutr 1975; 105:428–438.
80. Hoberman HD, Sims EAH, Peters JH. Creatine ans creatinine metabolism in the normal male adult studied with the aid of isotopic nitrogen. J Biol Chem 1948; 172:45–58.
81. Huso ME, Hampl JS, Johnston CS, Swan PD. Creatine supplementation influences substrate utilization at rest. J Appl Physiol 2002; 93:2018–2022.
82. Izquierdo M, Ibanez J, Gonzalez-Badillo JJ, Gorostiaga EM. Effects of creatine supplementation on muscle power, endurance, ans sprint performance. Med Sci Sport Exerc 2002; 34:332–343.
83. Mujika I, Padilla S, Ibanez J, Izquierdo M, Gorostiaga EM. Creatine supplementation and sprint performance in soccer players. Med Sci Sport Exerc 2000; 32:518–525.
84. Peyrebrune MC, Nevill ME, Donaldson FD, Cosford DJ. The effects of oral creatine supplementation on performance in single and repeated sprint swimming. J Sports Sci 1998; 16:271–279.
85. Rossiter HB. The effect of oral creatine supplementation on the 1000m performance of competitive rowers. J Sports Sci 1996; 14:175–179.
86. Poortmans JR, Auquier H, Renaut V, Durussel A, Saugy M, Brisson G. Effects of short-term creatine supplementation on renal responses in men. Eur J Appl Physiol 1997; 76:566, 567.
87. Rockwell JA, Rankin JW, Toderico B. Creatine supplementation affects muscle creatine during energy restriction. Med Sci Sport Exerc 2001; 33:61–68.
88. Volek JS, Mazzetti SA, Farquhar WB, Barnes BR, Gomez AL, Kraemer WJ. Physiological responses to short-term exercise in the heat after creatine loading. Med Sci Sport Exerc 2001; 33:1101–1108.

89. Havenetidis K, Bourdas D. Creatine supplementation: effects on urinary excretion and anaerobic performance. J Sports Med Phys Fitness 2003; 43:347–355.
90. Bermon S, Venembre P, Sachet C, Valour S, Dolisi C. Effects of creatine monohydrate ingestion in sedentary and weight-trained older adults. Acta Physiol Scand 1998; 164:147–155.
91. Maganaris C, Maughan R. Creatine supplementation enhances maximum volunbtary isometric force and endurance capacity in resistance trained men. Acta Physiol Scand 1998; 163:279–287.
92. Burke DG, Chilibeck PD, Davison KS, Candow DG, Farthing JP, Smith-Palmer T. The effect of wey protein supplementation with and without creatine monohydrate combined with resistance training on lean tissue mass and muscle strength. Int J Sport Nutr 2001; 11:349–364.
93. Harris RC, Soderlund K, Hultman E. Elevation of creatine in resting and exercised muscle of normal subjects by creatine supplementation. Clin Sci 1992; 83:367–374.
94. Kilduff LP, Vidakovic P, Cooney G, et al. Effects of creatine on isometric bench^press performance in resistance-trained humans. Med Sci Sport Exerc 2002; 34:1176–1183.
95. Poortmans JR, Kumps A, Duez P, Fofonka A, Carpentier A, Francaux M. Effect of oral creatine supplementation on urinary methylamine, formaldehyde, and formate. Med Sci Sport Exerc 2005; 37:1717–1720.
96. Rawson ES, Clarkson PM, Price TB, Miles MP. Differential response of muscle phosphocreatine to creatine supplementation in young and old subjects. Acta Physiol Scand 2002; 174:57–65.
97. Steenge GR, Simpson EJ, Greenhaff PL. Protein- and carbohydrate-induced augmentation of whole body creatine retention in humans. J Appl Physiol 2000; 89:1165–1171.
98. Vandenberghe K, Van Hecke P, Van Leemputte M, Vangerven L, Hespel P. Phosphocreatine resynthesis is not affected by creatine loading. Med Sci Sport Exerc 1999; 31:236–242.
99. Kuehl K, Goldberg L, Elliott D. Letter to the Editor-in-chief. Re: Long-term oral creatine supplementation does not impair renal function in healthy athletes. Med Sci Sports Exerc 2000; 32:248.
100. Kuehl K, Koehler S, Dulacki K, et al. Effects of oral creatine monohydrate supplementation on renal function in adults. Med Sci Sports Exerc 2000; 32:S168.
101. Kreider R, Ransom J, Rasmussen C, et al. Creatine supplementation during preseason football training does not affect markers of renal function. FASEB J 1999; 13:A543.
102. Kreider R, Rasmussen C, Melton C, et al. Long-term creatine supplementation does not adversely affect clinical markers of health. Med Sci Sports Exerc 2000; 32:S134.
103. Evans G, Greaves I. Microalbuminuria as predictor of outcome. Brit Med J 1999; 318:207, 208.
104. Camamori ML, Fioretto M. The need for early predictors of diabetic nephrpathy risk. Diabetes 2000; 49:1399–1408.
105. Mattock MB. Prospective study of microalbuminuria as predictor of mortality in NIDDM. Diabetes 1992; 41:736–741.
106. Mogensen CE. Prediction in clinical diabetic nephrpoathy in IDDM patients. Diabetes 1990; 39:761–767.

107. Groeneveld GJ, Beijer C, Veldink JH, Kalmijn S, Wokke JHJ, Van den Berg LH. Few adverse effects of long-term creatine supplementation in a placebo-controlled trial. Eur J Sports Med 2005; 26:307–313.
108. Edmunds JW, Jayapalan S, DiMarco NM, Saboorian MH, Aukema HM. Creatine supplementation increases renal disease progression in Han:SPRD-cy rats. Am J Kidney Dis 2001; 37:73–79.
109. Taes YEC, Delanghe JR, Wuyts B, Van de Voorde J, Lameire NH. Creatine supplementation does not affect kidney function in an animal model with pre-existing renal failure. Nephrol Dial Transplant 2003; 18:258–264.
110. Ferreira LG, Bergamaschi CT, Lazaretti-Castro M, Heilberg IP. Effects of creatine supplementation on body composition and renal function in rats. Med Sci Sport Exerc 2005; 37:1525–1529.
111. Koshy KM, Griswold E, Schneeberger EE. Interstitial nephritis in a patient taking creatine. New Engl J Med 1999; 340:814, 815.
112. Wyss M, Kaddurah-Daouk R. Creatine and creatinine metabolism. Physiol Rev 2000; 80(3):1107–1213.
113. Yu PH, Zuo DM. Formaldehyde produced endogenously via deamination of methylamine. A potential risk factor for initiation of endothelial injury. Atherosclerosis 1996; 120(1–2):189–197.
114. Deng Y, Boomsma F, Yu PH. Deamination of methylamine and aminoacetone increases aldehydes and oxidative stress in rats. Life Sci 1998; 63(23):2049–2058.
115. Mitchell SC, Zhang AQ. Methylamine in human urine. Clin Chim Acta 2001; 312(1–2):107–114.
116. Yu PH, Wright S, Fan EH, Lun ZR, Gubisne-Haberle D. Physiological and pathological implications of semicarbazide-sensitive amine oxidase. Biochim Biophys Acta 2003; 1647(1–2):193–199.
117. Headlam HA, Mortimer A, Easton CJ. Beta-scission of C-3 (beta carbon) alkoxyl radicals on peptides and proteins: a novel pathway which results in the formation of alpha-carbon radicals and the loss of amino acid side chains. Chem Res Toxicol 2000; 13:1087–1095.
118. Quievryn G, Zhitkovich A. Loss of DNA-protein crosslinks from formaldehyde-exposed cells occurs through spontaneous hydrolysis and an active repair process linked to proteasome function. Carcinogenesis 2000; 21:1573–1580.
119. Yu PH, Deng Y. Potential cytotoxic effect of chronic administration of creatine, a nutrition supplement to augment athletic performance. Med Hypotheses 2000; 54(5):726–728.
120. Kapeller-Adler R, Toda K. Uber das vorkommen von monomethylamin im harn. Biochem Z 1932; 248:403–425.
121. Boeniger MF. Formate in urine as a biological indicator of formaldehyde exposure: a review. Am Ind Hyg Assoc J 1987; 48(11):900–908.
122. Berode M, Sethre T, Laubli T, Savolainen H. Urinary methanol and formic acid as indicators of occupational exposure to methyl formate. Int Arch Occup Environ Health 2000; 73(6):410–414.
123. Kage S, Kudo K, Ikeda H, Ikeda N. Simultaneous determination of formate and acetate in whole blood and urine from humans using gas chromatography-mass spectrometry. J Chromatogr B Analyt Technol Biomed Life Sci 2004; 805(1):113–117.
124. Schmidt FH. [Faulty measurement of urinary glucose concentration by polarization]. Dtsch Med Wochenschr 1967; 92(44):2025–2027.

125. Yu PH, Zuo DM. Oxidative deamination of methylamine by semicarbazide-sensitive amine oxidase leads to cytotoxic damage in endothelial cells. Possible consequences for diabetes. Diabetes 1993; 42(4):594–603.
126. Garpenstrand H, Bergqvist M, Brattstrom D, et al. Serum semicarbazide-sensitive amine oxidase (SSAO) activity correlates with VEGF in non-small-cell lung cancer patients. Med Oncol 2004; 21(3):241–250.
127. Kinemuchi H, Sugimoto H, Obata T, Satoh N, Ueda S. Selective inhibitors of membrane-bound semicarbazide-sensitive amine oxidase (SSAO) activity in mammalian tissues. Neurotoxicology 2004; 25(1–2):325–335.
128. Remuzzi G, Weening JJ. Albuminuria as early test for vascular disease. Lancet 2005; 365(9459):556, 557.
129. Soffritti M, Belpoggi F, Lambertin L, Lauriola M, Padovani MM, Maltoni C. Results of long-term experimental studies on the carcinogenicity of formaldehyde and acetaldehyde in rats. Ann N Y Acad Sci 2002; 982:87–105.
130. Gooderham NJ, Murray S, Lynch AM, et al. Food-derived heterocyclic amine mutagens: Variable metabolism and significance to humans. Drug Metabol Disposition 2001; 29:529–534.
131. Knize MG, Salmon CP, Pais P, Felton JS. Food heating and the formation of heterocyclic aromatic amine and polycyclic aromatic hydrocarbon mutagens/carcinogens. Adv Exp Med Biol 1999; 459:179–193.
132. Heddle JA, Knize MG, Dawod D, Zhang XB. A test of the mutagenicity of cooked meats in vivo. Mutagenesis 2001; 16:103–107.
133. Wyss M, Schulze A. Health implications of creatine: Can oral creatine supplementation protect against neurological and atherosclerotic disease? Neurosience 2002; 112:243–260.
134. Derave W, Vanden Eede E, Hespel P, Carmella SG, Hecht DS. Oral creatine supplementation in humans does not elevate urinary excretion of the carcinogen N-nitrososarcosine. Nutrition 2006; 22:332, 333.
135. Friesen MD, Rothman N, Strickland PT. Concentration of 2-amino-1-methyl-6-phenylimidazo(4,5-b)pyridine (PhIP) in urine and alkali-hydrolyzedn urine after consumption of charbroiled beef. Cancer Lett 2001; 173:43–51.
136. Knize MG, Kulp KS, Malfatti MA, Salmon CP, Felton JS. Liquid chromatography-tandem mass spectrometry method of urine analysis for determining human variation in carcinogen metabolism. J Chromatogr 2001; 914:95–103.
137. Knize MG, Kulp KS, Salmon CP, Keating GA, Felton JS. Factors affecting human heterocyclic amine intake and the metabolism of PhIP. Mutation Res 2002; 9377:1–10.
138. Toribio F, Moyano E, Puignou L, Galceran MT. Ion-trap tandem mass spectrometry for the determination of heterocyclic amines in food. J Chromatogr 2002; 948:267–281.
139. Poortmans JR, Francaux M. Renal implications of exogenous creatine monohydrate supplementation. Am J Med Sports 2002; 4:212–216.
140. Balsom P, Ekblom B, Söderlund K, Sjödin B. Creatine supplementation and dynamic high-intensity intermittent exercise. Scand J Med Sci Sports 1993; 3:143–149.
141. Balsom P, Harridge S, Söderlund K, Sjödin B, Ekblom B. Creatine supplementaion per se does not enhance endurance exercise performance. Acta Physiol Scand 1993; 149:521–523.

142. Greenhaff P, Bodin K, Söderlund K, Hutman E. Effect of oral creatine supplementation on skeletal muscle phosphocreatine resynthesis. Am J Physiol 1994; 266:E724–E730.
143. Stroud M, Holliman D, Bell D, Green A, Macdonald I, Greenhaff P. Effect of oral creatine supplementation on respiratory gas exchange and blood lactate accumulation during steady-stade incremental treadmill exercise znd recovery in man. Clin Sci 1994; 87:707–710.
144. Balsom P, Söderlund K, Sjödin B, Ekblom B. Skeletal muscle metabolism during short duration high-intensity exercise: influence of creatine supplementation. Acta Physiol Scand 1995; 154:303–310.
145. Dawson B, Cutler M, Moody A, Lawrence S, Goodman C, Randall N. Effects of oral creatine loading on single and repeated maximal short sprints. Aust J Sci Med Sport 1995; 27:56–61.
146. Mujika I, Chatard J, Lacoste L, Barale F, Geyssant A. Creatine supplementation does not improve sprint performance in competitive swimmers. Med Sci Sports Exerc 1996; 28:1435–1441.
147. Vandenberghe K, Gillis N, Van Leemputte M, Van Hecke P, Vangerven L, Hespel P. Caffeine counteracts the ergogenic action of muscle creatine loading. J Appl Physiol 1996; 80(452–457).
148. Becque M, Lochmann J, Melrose D. Effect of creatine supplementation during strength training on 1-RM and body composition. Med Sci Sports Exerc 1997; 29:S146.
149. Godly A, Yates J. Effects of creatine supplementation on endurance cycling combined with short, high-intensity bouts. Med Sci Sports Exerc 1997; 29:S251.
150. Grindstaff P, Kreider R, Bishop R, et al. Effects of creatine supplementation on repetitive sprint performance and body composition in competitive swimmers. Int J Sports Nutr 1997; 7:330–346.
151. Hamilton-Ward K, Meyers M, Skelly W, Marley R, Saunders J. Effect of creatine supplementation on upper extremity anaerobic response in females. Med Sci Sports Exerc 1997; 29:S146.
152. Prevost M, Nelson A, Morris G. Creatine supplementation enhances intermittent work performance. Res Quart Exerc Sport 1997; 68:233–240.
153. Stout J, Echerson J, Nooman D, Moore G, Cullen D. The effects of a supplement designed to augment creatine uptake on exercise performance and fat free mass in football players. Med Sci Sports Exerc 1997; 29:S251.
154. Terrillion K, Kolkhorst F, Dolgener F, Joslyn S. The effect of creatine supplementation on two 700-m maximal running bouts. Int J Sport Nutr 1997; 7:138–143.
155. Ööpik V, Pääsuke M, Timpmann S, Medijainen L, Ereline J, Smirnova T. Effect of creatine supplementation during rapid body mass reduction on metabolism and isokinetic muscle performance capacity. Eur J Appl Physiol 1998; 78:83–92.
156. Snow R, McKenna M, Selig S, Kemp J, Stathis C, Zhao S. Effect of creatine supplementation on sprint exercise performance and muscle metabolism. J Appl Physiol 1998; 84:1667–1673.
157. Volek JS, Duncan ND, Mazzetti SA, et al. Performance and muscle fiber adaptations to creatine supplementation and heavy resistance training. Med Sci Sports Exerc 1999; 31:1147–1156.

158. Ööpik V, Timpmann S, Medijainen L. Metabolic effect of creatine supplementation with or without a concomitant reduction in body weight. J Sports Sci 1999; 17:560–561.
159. Urbanski RL, Loy SF, Vincent WJ, Yaspelkis BB, III. Creatine supplementation differentially affects maximal isometric strength and time to fatigue in large and small muscle groups. Int J Sports Nutr 1999; 9:136–145.
160. Becque MD, Lochmann JD, Melrose DR. Effects of oral creatine supplementation on muscular strength and body composition. Med Sci Sport Exerc 2000; 32:654–658.
161. Mihic S, MacDonald JR, McKenzie S, Tarnopolsky MA. Acute creatine loading increases fat-free mass, but does not affect blood pressure, plasma creatinine, or CK activity in men and women. Med Sci Sport Exerc 2000; 32:291–296.
162. Rico-Sanz J, Mendez Marco MT. Cretaine enhances oxygen uptake and performance during alterning intensity exercise. Med Sci Sport Exerc 2000; 32:379–385.
163. Shomrat A, Weinstein Y, Katz A. Effect of creratine feeding on maximal exercise performance in vegetarians. Eur J Appl Physiol 2000; 82:321–325.
164. Deutekom M, Beltman JG, de Ruiter CJ, de Koning JJ, de Haan A. No acute effects of short-term creatine supplementation on muscle properties and sprint performance. Eur J Appl Physiol 2000; 82:223–229.
165. Bennett T, Bathalon G, Armstrong DR, et al. Effect of creatine on performance of militarily relevant tasks and soldier health. Mil Med 2001; 166:996–1002.
166. Finn JP, Ebert TR, Withers RT, et al. Effect of creatine supplementation on metabolism and performance in humans during intermittent sprint cycling. Eur J Appl Physiol 2001; 84:238–243.
167. Skare O-C, Skadberg O, Wisnes AR. Cretaine supplementation improves sprint performance in male sprinters. Scand J Med Sci Sports 2001; 11:96–102.
168. Wilder N, Deivert RG, Hagerman F, Gilders R. The effects of low-dose creatine supplementation versus creatine loading in collegiate football players. J Athletic Training 2001; 36:124–129.
169. Ziegenfuss TN, Rogers M, Lowery L, et al. Effect of creatine loading on anaerobic performance and skeletal muscle volume in NCAA division I athletes. Nutrition 2002; 18:397–402.
170. Saab G, Marsh GD, Casselman MA, Thompson RT. Changes in human muscle transverse relaxation following short-term creatine supplementation. Exper Physiol 2002; 87:383–389.
171. van Loon LJC, Ooosterlaar AM, Hartgens F, Hesselink MKC, Snow RJ, Wagenmakers AJM. Effects of creatine loading and prolonged creatine supplementation on body composition, fuel selection, sprint and endurance performance un humans. Clin Sci 2003; 104:153–162.
172. Mendes RR, Pires I, Oliveira A, Tirapegui J. Effects of creatine supplementation on the performance and body composition of competitive swimmers. J Nutr Biochem 2004; 15:473–478.
173. Rosene JM, Whitman SA, Fogarty TD. A comparison of thermoregulation with creatine supplementation between the sexes in a thermoneutral environment. J Athletic Training 2004; 39:50–55.
174. McConell GK, Shinewell J, Stephens TJ, Stahis CG, Canny BJ, Snow RJ. Creatine supplementation reduces muscle inosine monophosphate during endurance exercise in humans. Med Sci Sport Exerc 2005; 37:2054–2061.

175. Peyrebrune MC, Stokes K, Hall GM, Nevill ME. Effect of creatine supplementation on training for competition in elite swimmers. Med Sci Sport Exerc 2005; 37:2140–2147.
176. Earnest C, Snell P, Rodriguez R, Almada A, Mitchel T. The effect of creatine monohydrate ingstion on anaerobic power indices, muscular strength and body composition. Acta Physiol Scand 1995; 153:207–209.
177. Thompson C, Kemp G, Sanderson A, et al. Effect of creatine on aerobic and anaerobic metabolism in skeletal muscle in swimmers. Br J Sports Med 1996; 30:222–225.
178. Goldberg P, Bechtel P. Effects of low dose creatine supplementation on strength, speed and power events by male athletes. Med Sci Sports Exerc 1997; 29:S251.
179. Kirksey K, Warren B, Stone MH, Stone MR, Johnson R. The effect of six weeks of creatine monohydrate supplementationin male and female track athletes. Med Sci Sport Exerc 1997; 29:S145.
180. Volek J, Kraemer W, Bush J, et al. Creatine supplementation enhances muscular performance during high-intensity resistance exercise. J Am Diet Assoc 1997; 97:765–770.
181. Leenders N, Sherman WM, Lamb DR, Nelson TE. Creatine supplementation and swimming performance. Int J Sports Nutr 1999; 9:251–262.
182. Stone MH, Sanborn K, Smith LL, et al. Effects of in-season (5 weeks) creatine and pyruvate supplementation on anaerobic preformance and body composition in American football players. Int J Sport Nutr 1999; 9:146–165.
183. Rawson ES, Wehnert ML, Clarkson PM. Effects of 30 days of creatine ingestion in older men. Eur J Appl Physiol 1999; 80:139–144.
184. Francaux M, Demeure R, Goudemant JF, Poortmans JR. Effect of exogenous creatine supplementation on muscle PCr metabolism. Int J Sports Med 2000; 21:1–7.
185. Burke DG, Silver S. Holt LE, Smith-Palmer T, Culligan CJ, Chilibeck PD. The effect of continuous low dose creatine supplementation on force, power and total work. Int J Sport Nutr Exerc Metabol 2000; 10:235–244.
186. Chrusch MJ, Chilibeck PD, Chad KE, Davison KS, Burke DG. Creatine supplementation combined with resistance training in older men. Med Sci Sport Exerc 2001; 33:2111–2117.
187. Brose A, Parise G, Tarnopolsky MA. Creatine supplementation enhances isometric and body composition improvements following strength exercise training in older adults. J Geront A Biol Sci Med Sci 2003; 58:11–19.
188. Eijnde BO, Van Leemputte M, Goris M, et al. Effects of creatine supplementation and exercise training on fitness in men 55-75 yr old. J Appl Physiol 2003; 95:818–828.
189. Eckerson JM, Bull AJ, Moore GA. The effect of 30 days of creatine phosphate supplementtion on body weight in men. Med Sci Sport Exerc 2003; 35:S217.
190. Chilibeck PD, Stride D, Farthing JP, Burke DG. Effect of creatine ingestion after exercise on muscle thickness in males and females. Med Sci Sport Exerc 2004; 36:1781–1788.
191. Volek JS, Ratamess NA, Rubin MR, et al. The effects of creatine supplementation on muscular performance and body composition responses to short-term resistance training overreaching. Eur J Appl Physiol 2004; 91:628–637.
192. Brose A, Parise G, Tarnopolsky MA. Creatine supplementation enhances isometric strength and body composition improvements following strength exercise training in older adults. J Gerontol Biol Sci 2003; 58:11–19.

193. Schröder H, Terrados N, Tramullas A. Risk assessment of the potential side effects of long-term creatine supplementation in team sport athletes. Eur J Nutr 2005; 44:255–261.
194. Sandhu RS, Como JJ, Scalea TS, Betts JM. Renal failure and exercise-induced rabdomyolysis in patients taking performance-enhancing compounds. J Trauma 2002; 53:761–763.
195. Haghighi M, Taylor WC. Effects of oral creatine on renal function. Med Sci Sport Exerc 2003; 35:S314.
196. Jones EC. Creatine, nephrolithiasis, and medullary sponge kidney. Med Sci Sport Exerc 2004; 36:S330.
197. Boswell L, Mistry D, Okusa M, et al. Creatine supplementation does not affect renal function at rest or during exercise. Med Sci Sport Exerc 2003; 35(Suppl) S400.

6 Clinical Applications

Joseph P. Weir, PhD

1. INTRODUCTION

Although creatine supplementation (CS) is typically considered in the context of sports supplementation, a continually expanding body of research literature is examining the potential clinical and therapeutic potential of CS. Aspects of clinical use seem obvious, to enhance muscle performance in conditions of sarcopenia, for muscular rehabilitation following injury, and for inborn errors of metabolism. In addition, the effects of CS have been examined in a variety of neurological and muscular disorders. Although most of the creatine (Cr) stored in the body is found in skeletal muscle, Cr is also found in tissues such as the brain, myocardial tissue, retina, and testes. Clinical application of CS targets these tissues as well as skeletal muscle.

2. GENERAL EFFECTS OF THE CR/CR PHOSPHATE/CR KINASE SYSTEM

2.1. Temporal Energy Buffer

The most straightforward clinical effect of CS would be its ability to enhance intracellular Cr and Cr phosphate (CrP) stores. The elevated CrP can then provide more substrate for the Cr kinase (CK) reaction, which would serve to help maintain ATP levels during periods of energetic stress, such as in ischemia or hypoxia. For example, Wilken et al. *(1)* have shown that Cr treatment attenuates the decline of ATP in mouse-brainstem slices exposed to hypoxia. The Cr/CrP/CK system serves to rapidly rephosphorylate ADP to ATP by donating a phosphate group to ADP. This is sometimes referred to as a temporal energy buffer

From: *Essentials of Creatine in Sports and Health*
Edited by: J. R. Stout, J. Antonio and D. Kalman © Humana Press Inc., Totowa, NJ

in the cell because over periods of several seconds of high level of energy consumption, the ADP and ATP concentrations remain fairly constant *(2,3)*. It is this temporal energy buffer function that is typically targeted in the use of CS as a sport supplement. The use of CS can increase intramuscular Cr and CrP, thus increasing the energy buffer capacity of the cell in a manner somewhat like charging a battery. The battery analogy may have clinical implications, especially with respect to the amelioration of ischemic damage.

However, the simple temporal energy buffer does not appear to explain the beneficial effects of Cr under all conditions. For example, Berger et al. *(4)* found the Cr treatment protected brain tissue in animal models of hypoxic injury, but the effects were independent of preservation of ATP levels. Further, CS has been shown to affect the levels of transcription factors that affect muscle growth *(5)*, suggesting that Cr may facilitate muscle growth through mechanisms beyond enhanced work capacity owing to augmented energy charge.

A variety of other effects of the Cr/CrP/CK system have clinical implications in this regard. Perhaps most importantly, Cr and CrP are intimately involved in the function of the mitochondria. Neurological diseases such as amyotrophic lateral sclerosis (ALS, Lou Gehrig's disease) and Huntington's disease, in which neuronal cell death occurs, appear to be influenced by mitochondrial dysfunction *(6)*. In addition, the pathophysiology of acute brain insults such as stroke and traumatic brain injury (TBI) are tied to mitochondrial dysfunction *(7)*.

2.2. Antioxidant

The mitochondria are the source of oxidative ATP production. The mitochondria are also a major source of reactive oxygen species (ROS) and reactive nitrogen species (RNS) *(8,9)* and are involved in Ca^{2+} homeostasis *(10)*. In addition, they play a central role in pathways of cell death. The relationships between mitochondria, the Cr/CrP/CK system, and cell death through apoptosis and necrosis will be examined in more detail in Section 3.

With respect to ROS and RNS, Lawler et al. *(11)* were the first to show that Cr has direct antioxidant properties. Their cell-free in vitro analysis showed that Cr acted as an "antioxidant scavenger" against radical ions such as peroxynitrite and superoxide anions. In contrast, Cr had no effect on hydrogen peroxide and lipid peroxidation. It was speculated that Cr may have other more indirect antioxidant effects such as increasing intracellular levels of arginine, also an antioxidant *(11)*. Because neuromuscular diseases such as ALS and Huntington's disease have been associated with oxidative stress, a potential neuroprotective benefit of CS might be its antioxidant properties. More recently, Sestili et al. *(12)* have shown that Cr has direct cellular antioxidant properties. Three different cell lines were challenged with three oxidants, including hydrogen peroxide. Cr treatment increased cell survival, in a dose-dependent manner, in all cell lines with all three oxidants. The antioxidant effects were likely independent of the effect of CS on cellular energy charge because one cell line contained virtually no CK, so that CS had virtually no effect on CrP, yet antioxidant activity and cell survival were still enhanced with Cr.

2.3. Spatial Energy Buffer

The Cr/CrP/CK system appears to serve as an energy shuttle between the mitochondria and the sites of energy (ATP) consumption *(2,3,9)*. Different isoforms of the CK enzyme are present in different tissues (e.g., brain vs muscle) and in different compartments (e.g., mitochondria vs cytosol). In addition, the organization of CK molecules differs between compartments: cytosolic CK exists as a dimer, whereas mitochondrial CK (mtCK) functions as an octomer *(9)*. Recall that mitochondria have an inner and outer membrane, and that the process of oxidative phosphorylation occurs across the inner mitochondrial membrane. The sites of ATP consumption, for example, myosin ATPase for muscle contraction, occur outside the mitochondria. In the CrP-shuttle hypothesis, the transport of ATP from the mitochondria to the ATPase enzymes does not exclusively occur by simple diffusion of ATP. Rather, the mtCK isoform, located in the space between the inner and outer mitochondrial membranes, catalyzes the reaction whereby ATP donates the terminal phosphate group to Cr; the resulting CrP can then diffuse to the sites of energy consumption (CrP diffuses somewhat more easily across the outer mitochondrial membrane than ATP *[9]*) where the cytosolic CK isoform catalyzes the transfer of phosphate to ADP, thus generating ATP at the site of consumption *(13)*. Free-Cr can then diffuse back to the mitochondria to repeat the process.

The mtCK isoform favors the reaction whereby ATP and Cr react to form CrP and ADP. As noted in the previous paragraph CrP can then diffuse to the sites of extramitochondrial ATP consumption (e.g., myosin ATPase). In addition, production of ADP through this reaction in the mitochondria serves as a stimulator of mitochondrial respiration *(14)*. Smith et al. *(15)* have shown in humans that indices of mitochondrial respiration following exercise are correlated with the concentrations of CrP and Cr at the end of the exercise bout, providing in vivo evidence, albeit correlational, that mitochondrial respiratory rates are influenced by CrP and Cr.

Table 1
Cellular Effects of Cr

Temporal energy buffer (CK reaction)
Antioxidant
Spatial energy buffer (energy shuttle from mitochondria to cytosol)
Stimulation of mitochondrial respiration
pH buffer
Membrane stabilization

2.4. pH Buffer

The Cr/CrP/CK system serves as a pH buffer. Under conditions of high ATP demand, as in high-intensity exercise, the CK reaction favors ATP production. This reaction consumes a proton, thus decreasing acidity. During recovery, the reaction favors rephosphorylation of Cr to CrP, releasing a proton and decreasing pH *(16)*.

2.5. Membrane Stabilization

CrP has areas of both positive (phosphate group) and negative charge; i.e., it is a zwitterion. CrP can then bind to the heads of phospholipids in biological membranes. This acts to stabilize these membranes by decreasing membrane fluidity, which increases membrane integrity and decreases leaking of intramembranous contents to the outside of the membrane *(16)* (Table 1).

3. CELL DEATH

There are two general processes of cell death, necrosis and apoptosis. Necrosis occurs directly as a result of cell injury such as severe ischemia (e.g., as in a stroke). In necrosis, the "abrupt biochemical collapse... leads to the generation of free-radicals and excitotoxins (e.g., glutamate, cytotoxic cytokines, and calcium)" *(17)*. Ultimately, nuclear and cytoplasmic membranes rupture, and enzymatic degradation of DNA occurs *(17)*. Apoptosis, or programmed cell death (sometimes referred to as cell suicide), is a process of cell death in which a trigger(s) initiates a cascade of biochemical events, centered around activation of a family of proteins referred to as the caspases. So called "upstream" caspases respond to triggers and ultimately activate "downsteam" caspases that finally kill the cell by degrading DNA *(17)*.

"Necrosis is a passive process, characterized by cell and organelle swelling with release of the intracellular contents into the extracellular space. This usually results in inflammatory reactions, vascular damage, edema, and injury to the surrounding tissue. Apoptosis is connected with cell shrinkage, organelle relocation and compaction, chromatin condensation, and DNA cleavage into large fragments" (18).

3.1. Mitochondrial Permeability Transition Pore

Mitochondria are a key feature in processes of cell death. Both apoptosis and necrosis can be mediated by loss of mitochondrial integrity. Mitochondrial integrity is facilitated by "contact sites" between the inner and outer mitochondrial membranes *(9)*. These contact sites consist of complexes of various individual structures that span either the inner or outer mitochondrial membranes. For example, the adenine translocator structure, which crosses the inner mitochondrial membrane, and voltage-dependent anion channels, which cross the outer mitochondrial membrane, can interact with other structures, like mtCK, to form a contact site. Collectively, this system is referred to as the mitochondrial permeability transition (MPT) pore (*see* Fig. 1).

Apoptosis can be initiated by the permeabilization of the outer mitochondrial membrane, which causes the release of proteins that are normally sequestered in the space between the inner and outer mitochondrial membranes *(19)*. These proteins, such as cytochrome-*c* (a component of the electron transport chain which generates ATP through oxidative phosphorylation), then trigger the apoptotic cascade when released into the cytosol (e.g., activation of caspase 3). The inner mitochondrial membrane can be permeabilized by activation of the MPT, which is influenced by Ca^{2+} and ROS *(9)*. Activation of the MPT then interferes with oxidative ATP production, among other effects, which contributes to cell death by necrosis *(19)*. In addition, MPT activation can lead to apoptosis *(9)* (*see* Fig. 2). With respect to Ca^{2+}, the mitochondria play an important role in intracellular Ca^{2+} regulation and can store large quantities of Ca^{2+} in the mitochondrial matrix *(20,21)*. During periods of ischemia and oxidative stress, Ca^{2+} accumulates in the cytosol and ultimately in the mitochondrial matrix. The Ca^{2+} overload under these conditions can trigger the MPT opening *(22)*.

The activity of mtCK has been shown to inhibit activation of the MPT, thus inhibiting cell death *(9)*. This appears to be due to the action of mtCK in maintaining elevated levels of ADP in the mitochondrial matrix, as matrix ADP inhibits the MPT *(9)*. Furthermore, O'Gorman et al. *(23)* have

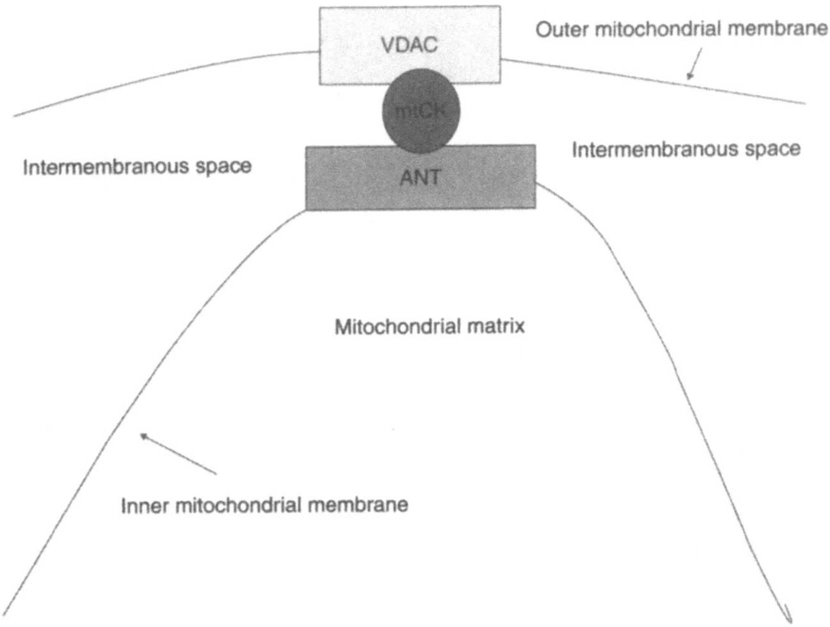

Fig. 1. Schematic representation of some components of the MPT pore. Under conditions of mitochondrial dysfunction, the interactions between the adenine translocator structure (ANT), mtCK, and voltage-dependent anion channels (VDAC) can be altered such that the MPT is opened, causing protein leak, diminished oxidative ATP production, and production of ROS. These effects can contribute to cell death by necrosis and apoptosis.

indicated that Cr helps stabilize mtCK in the octomeric form, which helps prevent activation of the MPT.

However, some recent studies have indicated that CS provides neuroprotection without evidence of effects on the MPT. Klivenyi et al. *(24)* showed that CS provided comparable neuroprotection against 1-methyl-4-phenyl-1-2-3-6-tetrahydropyridine (MPTP, a neurotoxin) administration in wild-type mice vs a transgenic mouse model that fails to express mtCK. Klivenyi et al. *(24)* argued that if the neuroprotection of CS is because of Cr interaction with the mtCK to inhibit opening of the MPT, then CS should not be neuroprotective in these transgenic mice. As CS did provide neuroprotection in both types of mice, their results suggest that the neuroprotective effects of CS are not because of effects on the MPT. Similarly, Brustovetsky et al. *(25)* reported that isolated mitochondria from Cr-fed rats (2% diet) did not exhibit a reduction in mitochondrial swelling in response to a Ca^{2+} challenge (a marker of MPT opening) relative to controls, suggesting that CS did not prevent Ca^{2+}-induced opening of the MPT.

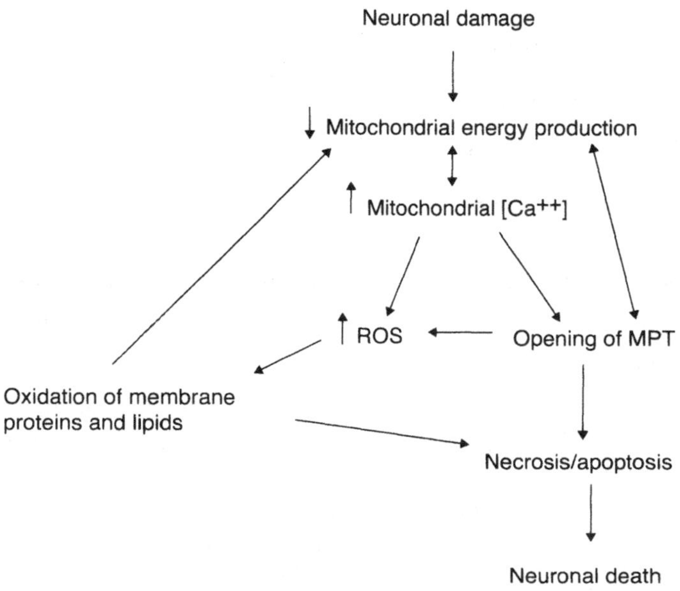

Fig. 2. Schematic representation of the role of mitochondria in cell death by necrosis and apoptosis emphasizing the role of the MPT pore.

As noted previously, the mitochondria produce ROS. CK is "extremely susceptible" to damage from ROS *(9)*. ROS and RNS can directly inhibit the enzymatic activity of CK. In addition, ROS stimulates the MPT and MPT in turn stimulates ROS formation. During conditions of ischemia-repurfusion injury (e.g., stroke), oxidative stress and Ca^{2+} contribute to the cellular damage. Interestingly, the damage primarily occurs during reperfusion when oxygen supply is resumed *(21)*.

3.2. Anoxic Depolarization

Balestrino et al. *(26)* suggest that neuronal cell death is tied to ATP, and therefore Cr, through the process of anoxic depolarization. In this model, when neurons experience anoxia, the membrane potential of the cell is maintained for a short period of time, during which ATP becomes depleted. As ATP levels fall, the Na^+/K^+-ATPase pumps fail, and loss of ionic gradients occurs. Specifically, Na^+, Cl^-, and probably most importantly Ca^{2+} enter the cell from the extracellular fluid, and K^+ is lost to the extracellular environment. The net change in ion concentrations results in cell membrane depolarization. Although short duration anoxic depolarization is reversible with reoxygenation, their work suggests that long-term anoxic depolarization is irreversible. However, high-dose Cr administration to slices of rat hippocampus undergoing anoxia markedly delayed anoxic

Table 2
Mitochondria and Cell Death

Decreased ATP synthesis
Apoptosis
ROS formation
Loss of Ca^{2+} homeostasis

depolarization *(26)*. The authors speculate that sufficiently high doses of Cr could be acutely administered to humans through parenteral injection.

3.3. Glutamate Excitotoxicity

Another key player in neurodegeneration is the process of glutamate excitotoxicity *(27)*. Glutamate is an amino acid that is also used as an excitatory neurotransmitter. Excess stimulation of glutamate receptors has been implicated in the pathophysiology of neurodegenerative diseases such as ALS and Huntington's disease, and in acute damage to the central nervous system such as in stroke and TBI *(18)*. Specifically, glutamate binds to the *N*-methyl-D-aspartate receptor (NMDA), which stimulates the movement of ions such as Ca^{2+} and Na^+ from the extracellular to intracellular space *(27,28)*. Excess glutamate stimulation (excitotoxicity) can lead to loss of Ca^{2+} homeostasis, leading to mitochondrial dysfunction, ROS formation, and stimulation of Ca^{2+}-activated proteases. Further, both Na^+ and Ca^{2+} dysregulation can lead to loss of membrane potential, osmotically mediated cell swelling, and cell lysis (*see* Fig. 3). The process of glutamate excitotoxicity can be due to decreased glutamate uptake by astroglial cells from the extracellular space (e.g., during ischemia), or in periods of compromised cellular energy, the neurons may be more sensitive to glutamate stimulation *(18)*.

Several studies have shown that Cr treatment appears protective against glutamate excitotoxicity. Xu et al. *(29)* have shown that CrP stimulates glutamate uptake, which might indicate that CS, and associated elevations in CrP, would help reduce extracellular glutamate. However, Cr also has a direct effect on cell death when cells are experimentally challenged with glutamate *(25,30–32)*. For example, Brewer and Wallimann *(31)* showed that Cr treatment protected embryonic rat neurons (from the hippocampus) that were challenged with glutamate. The authors attributed the protective effects of Cr, at least in part, to maintenance of CrP levels "which decline in the presence of glutamate without Cr" *(31)*. Other studies have also shown that Cr treatment increased survival of rat neurons challenged with glutamate *(25,30)*. Further, Malcon et al. *(32)* showed similar results in sections of brain tissue that were challenged with direct stimulation of NMDA receptors from rats fed a 1% Cr diet (Table 2).

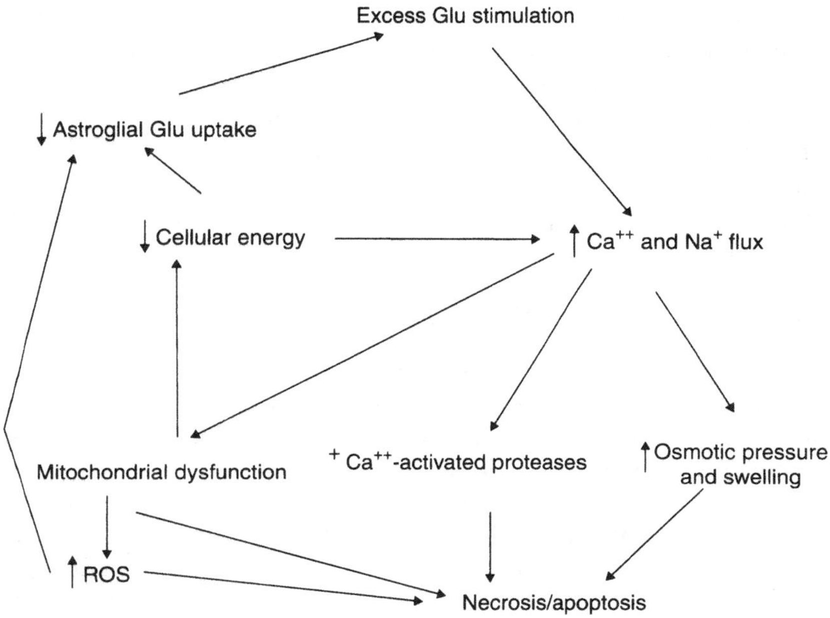

Fig. 3. Schematic representation of glutamate (Glu) excitotoxicity in cell death. The direct effect of glutamate on NMDA receptors is to increase ion flux across the cell membrane. In particular, loss of Ca^{2+} homeostasis can trigger a variety of cellular events that lead to cell death. In particular, Ca^{2+} accumulation in the mitochondria leads to mitochondrial dysfunction. This can directly contribute to cell death as described previously, and also feed back to further exacerbate the effect of glutamate.

"Levels of ATP, the primary energy source within cells, are carefully regulated in neurons. Impaired energy production results in the activation of excitatory amino acid receptors, increased intracellular Ca^{2+}, and the generation of free-radicals, which are events that are directly involved in necrosis and apoptosis. The key to maintaining the correct levels of ATP in the brain is the interaction between phosphocreatine and the enzyme CK" (33).

3.4. Variability in Response to Cr

However, to further complicate matters, recent studies suggest that CS may act in ways not directly tied to cellular energy charge and/or mitochondrial function. Pena-Altamira et al. *(34)* examined the effects of CS in wild-type rats and a transgenic mouse model of ALS. Wild-type rats were administered a neurotoxin in different brain regions. The benefits of CS (2% of diet) varied depending on brain region. The authors noted that "in addition to a general role of metabolic type, CS may have differential effects related to the nature, the brain topography, and the connectivity of different neuron types" *(34)*. They further argue that the typical hypotheses for the neuroprotective effects of CS (e.g., enhanced cellular energy charge, mitochondrial stability, and antioxidant effects) are difficult to reconcile with their results showing that, "different types of excitotoxic insults are ameliorated or not by Cr administration and why different neuronal populations respond or not to the dietary supplementation" *(34)*.

4. NEUROLOGICAL DISEASES

As a broad classification, neurodegenerative diseases can be considered acute or chronic diseases. Many acute (e.g., stroke, TBI, and spinal cord injury) and chronic neurodegenerative diseases (e.g., Parkinson's disease, Huntington's disease, and ALS) share the characteristic of neuronal cell death. In addition, the mechanisms of cell death, for example, apoptosis and necrosis are common between the diseases, even though the triggers for cell death may differ. In the following sections, a variety of studies, predominately in animal models, indicate that CS has potential as a therapeutic agent in acute neurological and neurodegenerative diseases.

4.1. Does CS Increase Brain-Cr Content?

A primary consideration regarding the utility of CS in neurological disorders is whether or not CS can increase brain-Cr content. Because Cr is a highly polar molecule, it does not readily cross the blood–brain barrier *(26)*, which potentially limits the usefulness of CS in treating injuries and diseases of the brain. Wilkinson et al. *(35)* did not find that short-term oral CS (5 g × four times per day × 5 d) increased white matter Cr content (magnetic resonance spectroscopy) in young male athletes. However, other studies have found that CS can increase brain-Cr content. Dechent et al. *(36)* have shown, using magnetic resonance spectroscopy, that oral CS (5 g/d × four times per day × 4 wk) in healthy volunteers can increase total-Cr content of human brain tissue, with the greatest effects shown in the white matter (increased 11.5%) and thalamus (14.6%), and more modest effects in gray matter (4.7%), and cerebellum (5.4%). These results have since been supported by Lyoo et al. *(37)*. Other data suggests that brain-Cr content can be modified in patients. Several case reports have been published showing that Cr deficiency in the brain can be corrected with CS *(38–41)*. Beyond case reports, Hersch et al. *(42)* found that 8 g/d × 16 wk of CS in individuals with Huntington's disease increased brain-Cr content by 13% in the occipital cortex and 75% in the frontal cortex (magnetic resonance spectroscopy). In situations wherein oral CS may not provide a sufficiently rapid response, Balestrino et al. *(26)* have suggested that parenteral injection of high doses of Cr might be useful to acutely increase brain-Cr content, for example, during periods of ischemia (e.g., stroke), although this has yet to be studied in detail.

4.2. Traumatic Brain Injury

Approximately 1.4 million Americans experience a TBI every year as a result of such events as automobile crashes *(43)*. Sports injuries account for approx 300,000 TBIs annually *(44)*. The effects of TBI can be categorized into primary and secondary factors. Primary factors are the direct mechanical result of the trauma to the brain at the time of injury. Secondary factors, which occur after the initial trauma, may result at least in part due to mitochondrial associated loss of Ca^{2+} homeostasis *(33)*.

Sullivan et al. *(33)* have shown, in an experimental model of TBI in mice and rats, that CS markedly reduced cortical lesion volume vs control animals. In addition, indices of mitochondrial function were higher in CS-treated animals. Specifically, the neuronal mitochondrial membrane potential was higher, ROS were lower, ATP concentration was higher, and Ca^{2+} concentration in the mitochondria was lower in CS-treated animals. In addition, Ca^{2+}-induced mitochondrial swelling, an index of activation of the MPT, was suppressed in CS-treated animals, but was elevated in control animals with TBI. The authors note that their results indicated that "…the neuroprotection afforded by Cr may involve preservation of normal levels of mitochondrial ATP, membrane potential, and calcium, conceivably through inhibition of the (MPT pore)." A subsequent study from the same group again showed that Cr-treated rats had smaller cortical lesions following experimentally induced TBI *(45)*. The TBI induced increases in cortical and hippocampal levels of free-fatty acids and lactate. Elevated free-fatty acids and lactate are markers of secondary brain injury; free-fatty acids reflect damage to the cell membrane (phospholipid breakdown—this can further lead to cellular damage through effects on blood flow, integrity of the blood–brain barrier, and free-radical formation) whereas lactate reflects altered cellular bioenergetics (and likely a decrease in pH). In addition to smaller lesion size, animals fed Cr had lower levels of free-fatty acids and lactate from the sites of injury *(45)*.

4.3. Stroke

In stroke, cell death by both necrosis and apoptosis occur. During a stroke, there is an "ischemic core" that consists of cells that were most acutely impacted by the decrease in blood supply; these cells die primarily by necrosis *(17)*. Surrounding the ischemic core is the "ischemic penumbra," which consists of cells that are less severely affected by the initial ischemia. In the penumbra, collateral circulation lessens the severity of the hypoxia and only cells that "reach a critical threshold of injury" will undergo apoptosis *(17)*. As cells that die in the penumbra are most likely to have been killed by apoptosis *(17)*, interventions that might lessen the apoptotic process are under study. In theory, CS might lessen the neuronal cell death in both the ischemic core and the penumbra through the ATP buffer function of the Cr/CrP/CK system.

Recent animal studies indicate that CS may indeed offer neuroprotection in stroke. Zhu et al. *(46)* examined the effect of CS in a mouse model of stroke. Mice fed Cr supplemented chow for 4 wk before experimental induction of cerebral ischemia (occlusion of middle cerebral artery for 2 h followed by 24 h of reperfusion) had smaller infarcts and experienced fewer motor neurological deficits than control mice. The Cr and ATP content in the brain on the side of ischemia were higher in the Cr-fed mice than controls. In addition, Cr-fed mice showed reduced levels of markers of apoptosis (cytochrome-*c* release and caspase-3 activation). Prass et al. *(47)* also showed that Cr feeding decreased infarct volume in a mouse model of stroke. However, their results showed that the Cr feeding had no effect on brain levels of Cr, CrP, ATP, or CK. Instead, the Cr treatment appeared to act by increasing vascular reactivity. For example, the middle cerebral artery from Cr-fed mice had a larger dilatory response to K+ and acidosis than control animals. In addition, cerebral blood flow during recovery following brain ischemia was enhanced in Cr-fed animals. Thus, it appears that CS may minimize brain damage that occurs with TBI and stroke, although no direct human data yet exist. The exact mechanisms of protection in animal models are as yet unclear.

4.4. Amyotrophic Lateral Sclerosis

ALS (also known as Lou Gehrig's disease) is a neuromuscular disease that results in the degeneration and death of both upper- and lower-motor neurons, resulting in muscle weakness, respiratory failure, and death typically within 5 yr after diagnosis *(48,49)*. Most cases of ALS have an unknown etiology, but a small percentage can be linked with a defect in the gene that codes for copper-zinc superoxide dismutase (SOD1), an enzyme that eliminates superoxide free-radicals. Thus, oxidative stress is implicated in the pathophysiology. In addition glutamate toxicity and mitochondrial abnormalities appear to contribute. There is no cure, but the drug riluzone, a glutamate antagonist, appears to have a small effect on survival *(49)*.

Several lines of evidence from animal models indicate that CS might have potential clinical utility in treating ALS. For example, Klivenyi et al. *(50)* have shown in a transgenic mouse model of ALS ("G93A" mice overexpress a mutant version of the SOD1 gene that results in symptoms of familial ALS) that Cr feeding increased survival time (1% Cr diet increase life-span by ~13 d relative to untreated controls; 2% increased life-span by ~26 d; in comparison riluzole increased life-span by ~13 d). Similarly, Andreassen et al. *(51)* found that Cr-treated G93A mice had improved survival, motor performance, and body weight compared with untreated animals (2% Cr diet had better results than either 1% or 3% Cr diets). Furthermore, ALS mice had elevated levels of brain glutamate, whereas CS reduced glutamate relative to untreated mice. The mechanism whereby CS can attenuate extracellular glutamate is unclear, but because glutamate uptake from the synapse is energetically demanding, the improved cellular energy charge might facilitate glutamate uptake. Snow et al. *(52)* also showed that CS increased survival time and delayed symptoms in transgenic mice (*SOD1* gene); however, the effects of Cr alone and riluzone alone were comparable with combined CS plus riluzone treatment (all groups delayed death by ~12 d relative to controls). The lack of additive effects of the combined therapy over single therapy suggested that both CS and riluzone may act through the same mechanism; possibly by inhibiting glutamate toxicity. Motor neuron survival was also significantly improved relative to control animals. Zhang et al. *(53)* examined the effects of CS and minocycline (an antibiotic found to inhibit MPT-mediated release of cytochrome-*c* from mitochondria to the cytosol, thus interfering with apoptosis) in a mouse model of ALS. They found that both CS and minocycline delayed ALS onset and mortality vs untreated mice. In addition, they found that combined treatment with Cr plus minocycline further delayed disease onset and mortality over either individual treatment. Although CS has been shown to significantly affect neuronal function in SOD1 mutant mice, it appears to have little to no effect on skeletal muscle function in this model *(54)*.

Unfortunately, trials in humans with ALS have been disappointing. Shefner et al. *(55)* conducted a clinical trial of CS in 104 subjects with ALS (51 subjects received CS of 20 g/d for 5 d followed by 5 g/d for 6 mo). There were no significant differences in the placebo and CS groups for any dependent variables, including muscle strength, motor unit number estimates, pulmonary function, and survival. Urinary Cr measures suggested that some of the placebo subjects acquired Cr outside the study protocol. Similarly, Groeneveld et al. *(56)* found no positive effect of CS (5 g twice daily) in patients with ALS ($n = 88$) relative to controls ($n = 87$) over a 16-mo trial. However, it has been suggested that these studies lacked sufficient statistical power and larger trials are needed *(57)*. Mazzini et al. *(58)* did show increases in isometric strength and muscle fatigue resistance after 7 d of CS (20 g/d); however, the lack of control/placebo group in the study does not allow one to be confident that this effect was not simply because of testing/practice effects. Despite the disappointing results in human trials, it has been suggested that larger doses that more closely approximate the animal studies are warranted *(34)*.

4.5. Spinal Cord Injury

Damage to the spinal cord in spinal cord injury includes both the initial mechanical trauma at the time of the event, but also secondary damage that occurs over time. This secondary damage includes the effects from edema, inflammation, and hypoxia. Hausmann et al. *(59)* have shown that Cr feeding 4 wk before and 4 d immediately following induction of traumatic spinal cord injury in rats resulted in a reduction in scar tissue formation and enhanced motor performance relative to control rats. These results suggest that CS might decrease the secondary damage associated with spinal cord injury. In contrast, Rabchevsky et al. *(60)* showed that the effect of CS in rat models of spinal cord injury did not affect motor performance, and the neurological effects depended on the nature of the injury. Specifically, more focal injuries appeared to respond better to CS than more diffuse injuries, and that in the focal injuries the beneficial effects were primarily located in the gray matter of the cord, as opposed to the white matter.

To date, three studies have examined CS in humans with spinal cord injury. These studies examined the muscular effects rather than the secondary effects on spinal cord tissue. Jacobs et al. *(61)* were the first to study CS in humans with spinal cord injury. They showed that in individuals with complete cervical spinal cord injuries at C5–C7, short-term CS (20 g/d for 7 d) improved maximal oxygen consumption during arm ergometery by approx 18% relative to placebo (double-blind placebo-controlled crossover design; washout period = 21 d, $n = 16$ males). However, two studies did not show a beneficial effect of CS. Kendall et al. *(62)* found that about 2 wk of oral CS (20 g/d for 6 d, 5 g/d thereafter) did not improve wrist extensor strength or endurance relative to placebo (randomized crossover design; 5-wk washout period) in nine individuals (one subject did not complete the trial because of loose stools) with incomplete tetraplegia (all subjects were at least 1-yr postinjury). Similarly, short-term CS (20 g/d for 6 d) did not improve performance on an 800 m wheelchair time trial relative to placebo (double-blind placebo crossover study; washout period = 28 d) in six trained wheelchair athletes (four with spinal cord injury). The duration of the time trial (~100 s) is such that glycolytic metabolism is the dominant energy system (blood lactate concentrations averaged 6–7 m*M*), which may have minimized the usefulness of CS for this task. Given the differences in exercise protocols and subject characteristics between these studies, generalizations on the effectiveness of CS in spinal cord injury are problematic. Nonetheless, given the data of Jacobs et al. *(61)*, which used a strong research design, the use of CS in spinal cord injury is worthy of further research.

4.6. Parkinson's Disease

Parkinson's disease stems from a decrease in dopamine production owing to loss of dopaminergic neurons in the brain, primarily in the basal ganglia. Motor symptoms in Parkinson's disease include tremor, rigidity, and bradykinesia. The pathophysiology appears, in part, to be tied to dysfunction of oxidative phosphorylation in mitochondria of at risk neurons *(63)*.

Parkinson's disease can be modeled in animals by administration of MPTP, which results in dysfunction of mitochondria (inhibited oxidative phosphorylation) in dopaminergic neurons. Specifically, a metabolite of MPTP called 1-methyl-4-phenylpyridinium (MPP+) is taken into dopaminergic cells by a dopamine transporter, where it accumulates in the mitochondria and inhibits oxidative phosphorylation *(64)*. Two studies have used this model to test the efficacy of CS. Matthews et al. *(65)* studied mice that were fed a Cr supplemented diet, a diet supplemented with a Cr analog (cyclocreatine), or placebo. The animals were then poisoned with MPTP. Cr treatment increased brain-Cr levels, decreased the MPTP-induced decline in dopamine and two dopamine metabolites, and minimized damage to dopaminergic neurons in the substantia nigra. Tests of motor function were not reported *(65)*. Similar results were found by Klivenyi et al. *(64)*.

To date, the only human data regarding CS and Parkinson's disease comes from a "futility trial" that examined CS (10 g/d), minocycline (an antimicrobial agent with anti-inflammatory effects), and placebo on the unified Parkinson's disease rating scale over 12 mo. A futility trial is performed as a screening study to eliminate agents that do not show promise for further study in subsequent large-scale (phase III) clinical trials. That is, agents are examined to determine if they are futile; futile agents are then unlikely to receive further consideration for inclusion in clinical trials. The results showed that Cr (and minocycline) "could not be rejected as futile…and therefore meet the criteria for consideration for further clinical testing *(63)*. It seems likely that a large-scale clinical trial for CS in Parkinson's disease will be forthcoming.

4.7. Huntington's Disease

Huntington's disease is a genetic (autosomal dominant) neurodegenerative disease that involves alterations in the protein called huntingtin. Specifically, the mutated version of huntingtin exhibits an exaggerated polyglutamine repeat. The length of the repeat varies and is correlated with rate of disease progression. The mutated protein undergoes proteolytic cleavage that results in peptide fragments, which can aggregate with other fragments as well as other proteins, resulting in large aggregates visible under microscopic examination of affected neurons. A primary site of neuronal damage is in the caudate and putamen of the basal ganglia, resulting in movement disorders such as chorea (jerky, involuntary movements) *(57)*. In addition, significant cognitive and behavioral symptoms occur. As the disease progresses, behavioral and cognitive symptoms worsen, and ultimately death ensues. Although the mechanistic link between huntingtin fragmentation and cell death is unclear, as with other neurodegenerative diseases, mitochondrial dysfunction appears to play a central role *(66)*. Indeed, a variety of studies have shown that energy metabolism is altered in brain and skeletal muscle, for example, elevated lactate concentration in the brain *(67)* and decreased ratio of CrP to inorganic phosphate (Pi) in skeletal muscle *(68)*. Furthermore, a transgenic mouse model of Huntington's disease has shown that both Cr and ATP concentrations are lower than in control animals *(66)*. Human data from magnetic resonance spectroscopy also shows decreased Cr levels in the basal ganglia in both asymptomatic individuals with the mutation, as well as in those with more advanced stages of the disease *(69)*. The magnitude of Cr deficit was correlated with the length of the polyglutamate repeat and with a clinical rating scale. Collectively, these studies indicate that Huntington's disease is tied to defects in cellular energy homeostasis, and suggest that Cr treatment may have clinical utility.

Studies of CS in animal models of Huntington's disease have been encouraging. There are two general animal models for Huntington's disease: (1) 3-nitropropionic acid (3-NP) administration, and (2) transgenic animals. With respect to 3-NP, this toxin inhibits the mitochondrial enzyme succinate dehydrogenase. The resulting energetic abnormality (ATP depletion, lactate accumulation) results in brain lesions and symptoms consistent with Huntington's disease. Matthews et al. *(70)* and Shear et al. *(71)* have both shown that Cr treatment decreases brain lesion size induced by 3-NP. In addition, Cr treatment minimized the effects of 3-NP on decreases in brain energetic substrates such as CrP, and ATP, and attenuated the increase in brain-lactate levels *(70)*. Furthermore, Cr-treated animals had smaller 3-NP-induced cognitive and motor deficits *(71)*. Similar results have been shown in 3-NP exposed mouse-brain slices *(72)*.

Ferrante et al. *(73)* used a transgenic mouse model (R6/2 model) of Huntington's disease and found that CS, started before the development of symptoms, delayed death, slowed brain atrophy, helped maintain body mass, and delayed deterioration of motor performance. A subsequent study by the same group showed that CS-delayed death, preserved motor performance, and maintained body mass relative to untreated controls, even when CS was delayed until after the development of symptoms (as might be the case in human therapy) *(66)*. However, better results were obtained the earlier the treatment was started. Similar results have been reported in another transgenic mouse model (N171-82Q) of Huntington's disease *(51)*.

As with ALS, the clinical trials in humans have been disappointing given the efficacy of CS in animal models. Verbessem et al. *(74)* studied the effects of CS on strength, cardiorespiratory endurance, and general motor function in 41 subjects with Huntington's disease ($n = 26$ with CS, $n = 15$ with placebo). One year of CS (5 g/d) failed to elicit any differences between those taking Cr vs those with placebo. Tabrizi et al. *(75,76)* performed an open label trial of CS (10 g/d) in a small group of individuals ($n = 13$) with Huntington's disease over the course of 2 yr. Because of the lack of control/placebo group, it is impossible to determine if CS was effective in slowing the progression of symptoms, but their results did show that the CS was well tolerated and magnetic resonance spectroscopy analyses showed that CS increased Cr levels in brain and muscle tissue. Bender et al. *(77)* have shown that 8–10 wk of oral CS (5 d at 20 g/d, then 6 g/d) in 20 individuals with Huntington's disease resulted in approx 20% reduction in Glx, a marker of glutamate and glutamine levels in the brain (magnetic resonance spectroscopy). Glutamate is an excitatory neurotransmitter and excess levels are neurotoxic and glutamate excitotoxicity is believed to play a role in the pathophysiology of Huntington's disease. Despite the apparent effect of CS on glutamate levels, there were no changes in brain-Cr levels or symptoms. Hersch et al. *(42)* reported that 16 wk of CS increased brain-Cr content and decreased levels of serum 8OH2'dG, a marker of DNA oxidative damage. The CS was well tolerated (no more adverse events than placebo). However, there were no differences in scores on the united Huntington's disease rating scale. A randomized placebo-controlled trial of CS in Huntington's disease (CREST-HD trial; http://clinicaltrials.gov/ct/show/NCT00026988) concluded data collection as of June, 2006; the results of this study have not been published as of this writing.

Thus, although animal survival and human biochemical responses suggest CS may be helpful in Huntington's disease, to date there is no evidence of clinical efficacy in humans. That said, Ryu et al. *(57)* have suggested that the human trials did not have enough statistical power "to be informative about whether or not Cr slows the clinical progression of HD;" however, they do attest to its safety and tolerability and favorable effects on serum and brain levels of Cr and on biomarkers of HD pathology." In addition, differences in dosage between animal and human studies should be noted. Verbessem et al. *(74)* have noted that typical animal studies (e.g., 2% Cr diet) result in doses approx 1.5 g/kg of body mass, whereas their human trial with a dose of 5 g/d resulted in a Cr load of approx 0.07 g/kg of body mass.

4.8. Multiple Sclerosis

Multiple sclerosis (MS) is an autoimmune neurological disease in which destruction of the myelin sheath in CNS neurons occurs. Individuals with MS experience a variety of signs and symptoms including fatigue, tremor, weakness, heat sensitivity, visual impairment, and poor balance and coordination. In addition, exercise capacity is compromised in MS and rephosphorylation of CrP is slowed *(78)*. To date, only one study has examined the effects of CS in MS. Lambert et al. *(79)* performed a double-blind placebo-controlled trial of CS (5 g Cr monohydrate × four times per day × 5 d) in 16 individuals (eight received CS, eight received placebo) with MS. CS had no significant effect on muscle performance (isokinetic fatigue of the thigh muscles at 180°/s). Muscle biopsies showed that the CS failed to significantly increase intramuscular concentrations of ATP, Cr, or CrP. Given the lack of CS effect on muscle-Cr levels, the failure of CS to increase exercise performance is not surprising. Future work should examine the mechanism(s) whereby a CS protocol shown to be effective in increasing Cr levels in healthy subjects fails to do so in those with MS. The authors speculated that Cr transporter levels might be compromised in MS, but there is no data in this regard.

4.9. Myasthenia Gravis

To date, no controlled studies have examined the effects of CS in myasthenia gravis. Stout et al. *(80)* have reported a case study of a male patient (age = 26 yr) with myasthenia gravis who engaged in a 15-wk program of resistance exercise and CS (5 g/d). The program resulted in marked increases in body mass (6.8%), fat-free mass (4.3%), and strength (12.5–37%). These increases occurred despite therapeutic use of prednisone (60 mg/d). In addition, the individual had previously engaged in resistance exercise after diagnosis, but had been unable to increase strength or body mass. The results of this case study suggest that a clinical trial examining the effects of Cr in conjunction with resistance exercise is warranted.

5. METABOLIC AND MUSCULAR DISEASES

5.1. Cr Deficiency Syndromes

Genetic diseases of Cr deficiency are thankfully rare, as the symptoms are severe (e.g., developmental delays, mental retardation, and seizures). However, several case reports have been published documenting Cr deficiency in the brain *(38–41)*. The first was published by Stockler et al. *(39)*, who reported a case of brain deficiency of Cr in a young child (22 mo). The child had diminished muscle tone, could not sit or roll over, and had difficulty swallowing. CS increased brain-Cr content (magnetic resonance spectroscopy) and improved symptoms *(40)*. One cause of Cr deficiency is a defect in the enzyme guanidinoacetate methyltransferase, which is the last step in the pathway of Cr biosysnthesis, and most case reports have identified this defect as the cause of the Cr deficiency. However, Cr deficiency in the brain may also be a result of dysfunction in Cr transport into the brain *(41)*. Regardless, all reports to date indicate that CS can increase brain-Cr levels and lessen symptoms *(38–41)*.

5.2. Mitochondrial Cytopathies

Mitochondrial cytopathies are a group of genetic diseases in which the mitochondria do not function properly. Most of these involve mutations in the mitochondrial DNA, but in some instances nuclear DNA can be involved *(81)*. If the dysfunction is primarily limited to skeletal muscle, the condition is termed a mitochondrial myopathy *(81)*. A compensatory adaptation to mitochondrial dysfunction is mitochondrial proliferation, which in skeletal muscle gives rise to the "ragged red fiber" when muscle samples are examined microscopically *(82)*. In addition, mitochondria also can exhibit paracrystalline inclusions, which are made of crystalized mtCK enzymes *(82)*. These inclusions, also known as mitochondrial inclusion bodies (MIBs) *(9)* are believed to reflect the upregulation of mtCK in an attempt to compensate for the effects of stress associated with the mitochondrial dysfunction *(9)*. The mtCK in MIBs does not appear to function properly *(9)*. Individuals with mitochondrial myopathies exhibit poor exercise tolerance, low VO_{2max}, and decreased peripheral O_2 extraction (a-VO_{2diff}) *(82)*.

Given the central role of the mitochondria in ATP production, mitochondrial cytopathies are prime targets for CS. To date relatively few studies have formally examined CS in mitochondrial cytopathies. A recent case report showed that CS eliminated paracrystalline inclusions in muscle biopsy samples from an endurance athlete with a mutation in the gene that codes for cytochrome-*b* *(83)*. Tarnopolsky et al. *(84)*

conducted a double-blind, placebo-controlled crossover study in seven individuals with mitochondrial cytopathies (primarily variants of MELAS: mitochondrial encephalopathy, lactic acidodis, and stroke-like episodes). Three weeks of CS (5 g Cr monohydrate two times per day for 2 wk, 2 g Cr monohydrate two times per day for 1 wk) resulted in significant increases in isometric handgrip and decreased dorsiflexor fatigue, but did not improve scores on activities of daily living, cycle ergometery, and the 2-min walk test. Postexercise lactate concentrations were higher with CS vs placebo. Borchert et al. *(85)* reported four cases (three were children) wherein patients with various mitochondrial cytopathies were treated using Cr at relatively high doses (0.1–0.2 g/Kg of body mass; about two to four times the dose used by Tarnopolsky et al. *[84]*) for 3 mo. They found improvements in maximal cycle ergometry power and submaximal cycling endurance. In contrast, Klopstock et al. *(86)* found no beneficial effects of CS in a double-blind, placebo-controlled crossover study with 16 individuals with mitochondrial dysfunction: 13 with chronic progressive external opthalmoplegia (CPEO) and three with mitochondrial myopathy. CPEO is a disease of mitochondrial DNA; because of the high levels of mitochondria in the extraocular muscle, symptoms of mitochondrial dysfunction can often present in these muscles. Differences in CS responses may be because of differences in the pathophysiology of MELAS vs CPEO. For example, resting CrP concentration in CPEO is not lower than in controls.

5.3. Muscular Dystrophy

Muscular dystrophy is a blanket term that describes several different specific genetic muscular diseases. All muscular dystrophies are characterized by progressive muscle wasting, weakness, and disability. Most of the muscular dystrophies are linked to defects in the dystrophin molecule in skeletal muscle. Dystrophin is a filamentous cytoskeletal protein that links the contractile protein actin to the basement membrane through the dystrophin–glycoprotein complex. The dystrophin–glycoprotein complex has both structural and cell signaling functions. Duchenne muscular dystrophy is the most common form of muscular dystrophy, and the most severe. Individuals with Duchenne muscular dystrophy lack the dystrophin protein, and intense muscle contractions, especially eccentric contractions, lead to excess muscle damage, and net muscle degeneration occurs over time. Much of the muscle damage is linked to loss of Ca^{2+} homeostasis, specifically increased Ca^{2+} concentration in the sarcoplasm, owing to greater Ca^{2+} "leak" resulting from instability in the sarcolemma *(87)*. The loss of Ca^{2+} homeostasis leads to mitochondrial dysfunction, activation of Ca^{2+}-activated proteases, and further decreases the ability of dystrophic muscle cells to process Ca^{2+}. Muscle damage leads to elevated levels of CK in the blood, as the protein leaks out of the muscle cells owing to loss of sarcolemmal integrity *(88)*. In addition, muscle levels of Cr and CrP are depressed *(88,89)*. There is no cure, and death usually occurs in the teenage years. The primary cause of death is respiratory failure because of weakness of the respiratory muscles. The only drug therapy known to slow progression of Duchenne muscular dystrophy is corticosteroid therapy; however, the side effects are significant. Becker muscular dystrophy is less common and less severe, as the dystrophin molecule is expressed, but an error in the dystrophin gene makes the dystrophin protein less effective. Individuals with Becker muscular dystrophy often survive to middle age. Other types of muscular dystrophy include fascioscapulohumeral, limb girdle, and oculopharyngeal muscular dystrophy.

A transgenic mouse model of Duchenne muscular dystrophy, the mdx mouse, has been developed. In the mdx mouse, dystrophin is not expressed because of a single point mutation in the dystrophin gene. Pulido et al. *(87)* examined the effects of Cr treatment in myotubes derived from mdx mice on the Ca^{2+} response to stressors known to increase Ca^{2+} concentration. Cr treatment attenuated the increase in Ca^{2+} concentration in mdx myotubes relative to controls. In addition, the mdx myotubes had lower levels of CrP than control myotubes and Cr treatment increased myotube-CrP concentration. Finally, the Cr treatment also increased mdx cell survival. Based on these results, the authors stated that "Cr is a substance with potential therapeutic properties which should be investigated as a possible adjuvant to established therapies" *(87)*. Passaquin et al. *(90)* examined Cr treatment in mdx pups. Animals were Cr treated through supplementation of their mothers' diets either at birth or 4 wk after birth. Because mdx mice experience a cycle of degeneration and regeneration at about 14–28 d of age, the treatments intervened before this first cycle. Both feeding protocols resulted in significant attenuation of necrosis relative to untreated mdx mice (>60% reduction in necrosis). However, these effects were limited to the primarily fast-twitch extensor digitorum longus, as the effects were not present in the primarily slow twitch soleus. In addition, indices of mitochondrial respiration were improved. Louis et al. *(89)* reported that Cr feeding in older (~3-mo old; "after the first wave of degeneration had occurred") mdx mice increased muscle-Cr content, had a small, although not statistically significant (~9%, $p = 0.08$) effect on isometric force, but did not affect other indices of muscle function such as responses to eccentric contractions (mdx muscle lost force during stretching contractions ~10 times faster than control muscle), skeletal muscle Ca^{2+} content, and centralization of nuclei (a marker of degeneration and subsequent regeneration). Thus, in these animal studies, the timing of Cr treatment might be important.

Human trials of CS in muscular dystrophy have been somewhat encouraging. Walter et al. *(91)* found that 8 wk of CS provided modest strength benefits in a study of subjects with a variety of different types of muscular dystrophy. Their double-blind, placebo-controlled crossover design showed that 10 g/d in adults, or 5 g/d in children, resulted in about 3% improvement in strength and ~10% improvement in a scale assessing activities of daily living. Similarly, Louis et al. *(92)* found that CS (3 g/d for 3 mo) markedly increased muscle strength (15%) and muscle endurance (approx two times) relative to placebo (double-blind crossover design with 2-mo washout period). In addition, bone mineral density and joint stiffness were improved. Interestingly, these increases occurred without concomitant increases in muscle CrP/ATP ratio, suggesting that "Cr supplementation may act through a mechanism other than energetic status of the cell" *(92)*. Tarnopolsky et al. *(93)* studied the effect of CS in 30 boys (mean age ~10 yr) with Duchenne muscular dystrophy. This double-blind, placebo-controlled crossover design (4 mo of CS or placebo, 6 wk of washout, then 4 mo of CS or placebo) showed that CS (2–5 g/d based on body size; dose ~0.1 g/kg/d) resulted in improved handgrip strength and increased fat-free mass relative to placebo; however, there were no significant Cr effects on pulmonary function or functional tests (e.g., walking time). Escolar et al. *(94)* did not find significant effects for CS (5 g/d, $n = 15$) relative to placebo ($n = 16$) or an alternative therapy (glutamine, $n = 19$) on strength, functional activities, or pulmonary function in boys (mean age ~6.5 yr) with Duchenne muscular dystrophy; however, they note that the trial lacked sufficient statistical power to detect the effects.

Another variant of muscular dystrophy is myotonic dystrophy. There are variants of myotonic dystrophy, but all have widespread effects beyond skeletal muscle. The genetic aspects differ from Duchenne and Becker muscular dystrophies as the coding errors occur in genes other than the dystrophin gene. For example, the most common variant of myotonic dystrophy, myotonic dystrophy type 1 (DM1; the most common adult genetic muscle disease), involves the gene dystrophia myotonica-protein kinase *(95)*. Symptoms include muscle atrophy, weakness, and excess muscle rigidity/spasms *(96)*. Two trials have examined the effects of CS on DM1 *(97,98)*. Both studies failed to show a statistically significant benefit of CS on measures of strength and function, although the paper by Walter et al. *(98)* appears to be underpowered given their statistical approach. Furthermore, the lack of response to CS may have been because of the inability of CS to increase muscle-Cr levels, as the magnetic spectroscopy data of Tarnopolsky showed no change in CrP/ATP ratio following CS.

Although the results in human studies of CS in muscular dystrophy are mixed, collectively the studies suggest that CS might be an effective adjuvant therapy to slow the deleterious effects of Duchenne and Becker muscular dystrophy. Given that pulmonary failure resulting from weakness of the respiratory muscles is one of the most common causes of death in Duchene muscular dystrophy *(99)*, treatments that delay the decline in respiratory muscle performance might increase longevity. To date, the trials of CS in muscular dystrophy do not show effects on pulmonary function; benefits in performance of other skeletal muscles indicate long-term clinical trials with larger sample sizes seem warranted.

5.4. McArdle's Disease

McArdle's disease is a genetic disease of muscular energy metabolism in which mutations in the gene that codes for the enzyme glycogen phosphorylase-b occur. The result is that individuals with McArdle's disease cannot engage in muscle glycogen breakdown, and therefore glycolytic metabolism is effectively blocked. During activity, individuals with McArdle's disease experience pronounced muscle fatigue, pain, and cramping *(100)*. In addition, they fail to generate lactate or develop acidosis during exercise *(82)*. Vorgerg et al. *(101)* found that CS (150 mg/kg/d for 1 wk followed by 60 mg/kg/d for the following 4 wk) produced modest changes in some measures of muscle performance (e.g., improvements in the force-time integral and spectral changes in the surface electromyographic signal during ischemic isometric exercise) but did not significantly improve performance in cycling or nonischemic isometric exercise. Interestingly, CS did not appear to increase muscle-CrP content (also recently shown by Zange et al. *[102]*) but did increase CrP depletion and accumulation of Pi, which suggests that "the beneficial effect was not due to an increase in energy availability but instead to an effect that somehow permitted exercise to continue to a level of greater energy depletion" *(100)*. It should also be noted that the small sample size (total $n = 9$) limited statistical power. A subsequent study by Vorgerd et al. *(103)*, using a comparable design, showed that higher dosages of Cr (150 mg/kg/d) actually worsened the symptoms of exercise intolerance. In addition, this study also showed that CS failed to increase CrP content, but unlike the data from CS at a lower dosage, there were no differences in CrP depletion and Pi accumulation. Tarnopolsky *(82)* have suggested that these studies collectively "...suggest that lower dosages (~4–5 g/d) may be of some benefit, whereas higher dosages (>10 g/d) may actually be deleterious to these patients." The inability of CS to increase CrP in McArdle's disease requires further study, as methods to facilitate Cr uptake in McArdle's disease may enhance the clinical utility of CS.

5.5. Gyrate Atrophy

Gyrate atrophy of the choroid and retina was the first disease to be treated with CS *(104)*. Gyrate atrophy is a genetic disease clinically expressed as progressive vision loss (diminished visual field, night blindness, cataract formation, myopia, and ultimately blindness). In addition, type II skeletal muscle fibers exhibit tubular aggregation and atrophy. Exercise intolerance is not typically reported in the literature with gyrate atrophy, but to date no rigorous studies evaluating exercise performance have been performed.

The biochemical defect involves the primary enzyme in ornithine catabolism, which results in high ornithine concentrations. Ornithine inhibits L-arginine-glycine amidinotransferase, which is the rate limiting step in Cr synthesis *(104–106)*. Individuals with gyrate atrophy also have diminished levels of Cr in the urine, muscle, and brain. Early papers examining gyrate atrophy had no control groups, so the results reflect weak evidence. In addition, the Cr dose of 1.5 g/d is well below the typical dose used in other studies of CS. Nonetheless, the data suggested little benefit for the visual symptoms, but muscle biopsy data indicated that CS helped reverse type II fiber atrophy *(104,107)*. More recently, it was shown that individuals with gyrate atrophy who engaged in long-term CS (1.5–2 g/d for >8 yr) had muscle-CrP/ATP ratios similar to those without gyrate atrophy, whereas individuals with untreated gyrate atrophy had values approx 70% of those taking Cr *(105)*. A study of brain-Cr levels in the same data set showed lowered Cr levels in individuals with gyrate atrophy that was partially corrected with long-term CS *(106)*. In addition, evidence of similar levels of brain atrophy was present in CS and nonCS subjects relative to those without gyrate atrophy.

Collectively, the available evidence suggests that CS can correct Cr levels in muscle, and to lesser extent in the brain, in those with gyrate atrophy. Although to date there is no evidence that the CS affects the visual symptoms, the studies are weakened by low doses of Cr. Future studies need to incorporate randomized controlled design, and use higher dosages to more fully evaluate the effect of CS on vision in gyrate atrophy.

6. CARDIOPULMONARY DISEASES

6.1. Heart Failure

Heart failure is characterized in large part by exercise intolerance *(108)*. The exercise intolerance is owing to both the dysfunction of the myocardium (maximal cardiac output ~50% vs controls) as well as decreased peripheral blood flow (diminished peripheral vasodilatory capacity) and skeletal muscle performance (especially reduced oxidative capacity and increased reliance on glycolytic metabolism) *(108)*.

Individuals with heart failure have also been shown to have decreased concentrations of Cr in both myocardial cells (~50%) *(109–111)* and skeletal muscle cells *(112)*. The severity of heart failure has been shown to be more severe in those with lower myocardial-Cr levels *(110,111)*. For example, the amount of Cr in myocardial cells of individuals with heart failure is correlated with left ventricular ejection fraction, indicating the severity of heart failure is related to the degree of Cr depletion *(113)*. Furthermore, the amount of Cr depletion is more severe in dilated cardiomyopathy than in hypertrophic cardiomyopathy *(113)*. Because Cr is not synthesized in myocardial cells, Cr content is determined by Cr transport from the blood. Neubauer et al. *(109)* have shown that the amount of Cr transporter in myocardial cells is reduced approx 30% in failing hearts from transplant recipients relative to the transplanted hearts. In addition to decreased Cr and Cr-transporter levels, the amount of ATP production in myocardial cells through the CK system is significantly compromised in heart failure such that the reduction in ATP synthesis by CK will likely "contribute to the pathophysiology of heart failure" *(112)*.

Gordon et al. *(114)* were the first to examine the effects of CS on cardiac and muscle performance in heart failure. In a double-blind placebo-controlled design, subjects with heart failure received either placebo ($n = 8$) or CS ($n = 9$; 20 g Cr/d) for 10 d. The CS significantly increased muscle concentrations (muscle biopsy) of Cr, CrP, and total-Cr. The CS had no significant effect on cardiac performance (resting ejection fraction), but significantly improved muscle exercise performance relative to placebo (one and two-legged cycle ergometery, isokinetic strength). Andrews et al. *(115)* also found that short term CS (5 g × four times per day for 5 d) improved skeletal muscle endurance performance (isometric forearm contractions at 75% of maximal voluntary contraction) in individuals with heart failure. Cardiac performance was not assessed. Given the effects of heart failure on skeletal muscle characteristics *(116)* and the importance of muscle fatigue on exercise capacity in heart failure, these results suggest that CS be a useful component in the treatment of exercise limitations in heart failure.

In contrast, Kuethe et al. *(117)* did not find a positive effect of CS in cardiac or endurance exercise performance (peak VO_2, ventilatory threshold, ejection fraction, and 6-min walk test) in 20 individuals with heart failure ($n = 13$ completed the double-blind, placebo-controlled, crossover trial; CS = 5 g × four times per day × 6 wk). However, both body weight (~2%) and forearm flexor muscle strength (~27%) were increased.

6.2. Chronic Obstructive Pulmonary Disease

Chronic obstructive pulmonary disease (COPD) is a relatively common and often debilitating pulmonary disease. Besides the pulmonary effects, COPD is also associated with muscle wasting and weakness. In addition, skeletal muscle from untrained individuals with COPD has been shown to exhibit altered cellular bioenergetics (e.g., higher Pi/CrP ratio, lower resting pH) *(118)*. Therefore, CS might benefit individuals with COPD independent of direct pulmonary effects. To date, only one study has examined CS in COPD *(119)*. Short-term CS (~15 g/d for 2 wk) increased muscle strength, muscle endurance, and fat-free mass relative to placebo. Subsequent long-term CS (5 g/d for 10 wk) in conjunction with a pulmonary rehabilitation program (including both endurance exercise and circuit weight training) showed further gains in strength and fat-free mass than placebo. Measures of quality of life were also increased with CS relative to placebo. Neither short- nor long-term CS improved measures of whole body endurance (walking tests, cycle ergometery) or pulmonary function. These limited results are encouraging with respect to management of COPD. Long-term studies are warranted.

7. REHABILITATION

Restoration and/or preservation of muscle mass and strength is central to rehabilitation following injury or disuse. In young healthy subjects, a variety of studies have shown that CS facilitates the development of muscle mass and strength when combined with progressive resistance exercise (for review *see* ref. *120*). It is logical then to examine the potential benefits of CS in dealing with the loss of strength and muscle mass in the rehabilitation process following injury.

Tyler et al. *(121)* found that CS (5 g/d) had no beneficial effect on the rehabilitation process (12 wk) following anterior cruciate ligament reconstruction surgery. However, the details of the strength training and rehabilitation process were limited, so it is somewhat difficult to directly compare these results with those of other studies. In contrast,

Hespel et al. *(5)* studied the effects of CS on recovery from muscle atrophy after forced immobilization. Healthy volunteers had their right thigh and leg placed in a cast for 2 wk. Half of the subjects were given Cr (20 g/d during 2 wk of immobilization, 15 g/d for 3 wk, then 5 g/d for 7 wk) whereas the other half received placebo. All subjects were given a strength training protocol for rehabilitation after the immobilization. Immobilization decreased muscle size and performance similarly in both groups. However, the Cr-treated subjects had a faster rate of recovery of muscle cross-sectional area and muscle power than placebo. Isometric strength changes were comparable between groups. Muscle biopsy data indicated that CS affected the expression of two myogenic transcription factors (these regulate muscle-specific gene expression) differently than placebo; specifically, levels of myogenin were lower after 10 wk of rehabilitation with CS, whereas levels of MRF4 were higher. Furthermore, the changes in muscle fiber area during rehabilitation were positively correlated with the levels of MRF4 ($r = 0.73$).

With respect to myogenic transcription factors, Willoughby et al. *(122)* have also shown that CS combined with resistance training affects levels of myogenic transcription factors relative to placebo, although in contrast to the data of Hespel et al. *(5)*, they showed CS increased levels of myogenin. These data suggest that CS facilitates muscle hypertrophy during rehabilitation and resistance training, and that some of this effect might be because of differences in regulation of gene expression. However, the influence of various myogenic transcription factors on muscle hypertrophy and atrophy, and the influence of Cr, requires further study.

8. AGING

A variety of studies have examined the short-term effects of CS in older adults *(123–125)*. In general, responses in older adults appear to be less robust than in younger subjects *(123,125)*. For example, CS was shown to result in smaller increases in muscle CrP in older (mean age ~70 yr) than younger (mean age ~20 yr) subjects *(124)*. However, a well-controlled study (3-wk familiarization period to control for learning effects of testing) by Gotshalk et al. *(126)* did show that 1 wk of CS (0.3 g/kg/d) significantly improved both upper and lower body muscular performance.

Given the importance of counteracting sarcopenia in older adults, herein the focus will be on long-term studies examining the combined effect of CS with resistance training. As with the rehabilitation data, studies examining the efficacy of CS in aging are mixed, with some studies showing a beneficial effect *(127,128)* and others showing no benefit *(129,130)*. Chrusch et al. *(128)* found that, in older adults (mean age ~70 yr), 12 wk of CS (~26 g/d for 5 d, then ~6 g/d thereafter; dosage was based on body size; $n = 16$) combined with resistance training resulted in significantly greater gains in lower body strength, lower body endurance, and fat-free mass than resistance training with a placebo ($n = 14$). Later analysis of bone mineral content data from this study showed that bone mineral content of the arms (but not the legs), was also increased in the Cr-treated subjects relative to placebo. The changes in arm bone mineral content were correlated with changes in lean tissue mass, which the authors interpreted to indicate that "…bone mineral content may be related to increases in muscles mass, which would increase muscle pull and strain on bone, providing stimulus for bone formation" *(131)*. A subsequent study showed that gains in strength and alterations in body composition in response to strength training (three times per week for 14 wk; load ~80% of the one-repetition maximum) in older adults (age >65 yr) were enhanced with CS (5 g/d) relative to placebo *(127)*. Increases in muscle total-Cr (~26%) were confirmed by muscle biopsy.

In contrast, Bermon et al. *(130)* reported that CS (20 g/d for 5 d, 5 g/d thereafter) for 8 wk, in conjunction with resistance exercise (three times per week; ~80% of one repetition maximum) failed to augment strength gains above those seen with placebo + resistance training in older adults (67–80 yr; both males and females; $n = 8$ per group). In a much longer-term study, Eijnde et al. *(129)* found that CS (5 g/d; up to 1 yr of supplementation), in conjunction with a combined cardiovascular and resistance training exercise program, had no beneficial effect on strength or cardiovascular endurance vs placebo in older men (55–75 yr). With respect to the strength data, it is important to note that the level of resistance was quite low (reps = 20–30/set). In addition, levels of muscle Cr, CrP, and ATP before the supplementation were higher than is typically reported for younger males, suggesting that CS had a limited potential to augment Cr stores.

Although it is problematic to make definitive conclusions regarding the differences between studies, two methodological aspects of the two negative trials deserve discussion. First, for the trial by Berman et al. *(130)*, the training protocol was of a short duration (8 wk) and the sample size was limited ($n = 8$ per group). In contrast, the positive trials by Chrusch et al. *(128)* and Brose et al. *(127)* were of 12 and 14 wk, respectively, and had sample sizes of 14–16 per group. Second, the low resistance used in the trial by Eijnde et al. *(129)* may have limited the utility of CS, as the duration of each set of 20–30 repetitions would likely minimize the importance of the phosphagen system. In contrast, the positive trials reported earlier had loads that limited repetitions to 10–12 per set. Finally, there is large interindividual variability in responses to Cr, which is tied in part to the Cr stores of the individual before supplementation. For example, it has been shown that those with lower levels of stored Cr before CS tend to respond better to CS than those with higher Cr levels before CS, for example, vegetarians tend to respond better to CS than meat eaters because the vegetarian diet is relatively low in Cr, which results in lower resting levels muscle total-Cr *(132)*. Sources of variability in the data (e.g., presupplement Cr levels) such as these add "noise" to the data and can mask the effects of CS.

9. SAFETY

A central issue with the use of CS is safety. There have been anecdotal reports in the press of safety concerns regarding muscle cramping and heat intolerance. In addition, Cr undergoes nonenzymatic conversion to creatinine, which is expelled in the urine. Therefore, an elevated urinary creatinine level is common with CS. Because urinary creatinine is used as a clinical marker of renal function, elevated urinary creatinine might be misinterpreted as evidence of renal damage. However, all of the studies cited earlier have failed to report any serious side effects. The most common complaints appear to concern gastrointestinal distress and loose stools. Moreover, Shao and Hathcock *(133)* have recently examined safety and dosing data from nearly 70 trials using CS. They stated that "…none of the clinical trials found a clear adverse effect related to Cr administration..." *(133)*. In addition, using standard procedures for assessing dosage levels from the literature, they determined that data "from well-designed randomized, controlled human trials indicates that the upper level for supplements for Cr is 5 g/d" *(133)*.

10. SUMMARY AND CONCLUSIONS

Although most of the attention on CS has focused on its use as a sports supplement, it is clear that there is potential for the use of Cr as a clinical agent. The clinical effects of Cr appear tied to mitochondrial function, including but not limited to enhancing cellular energy charge. There are two general clinical uses of Cr. First CS can aid skeletal muscle performance in a manner that is entirely analogous to its use as a sports supplement. For example, Cr might aid resistance training programs in the elderly. Second, Cr can affect neurological tissue to help prevent cell degeneration and cell death. Unfortunately, animal studies exhibit better efficacy than do human trials (although this pattern is not unique to Cr), and in some cases no clinical trials have yet been performed (e.g., stroke, TBI). To date, all the available literature indicates that use of Cr is safe, and only minor side effects are noted. It is also important to note that Cr use does not "cure" any disease. Rather, it may at best serve as an adjunct to standard clinical treatment.

11. PRACTICAL APPLICATIONS

Despite the volume of literature cited in this chapter, the clinical utility of CS is still open to more questions than answers. Therapists and nutritionists working with clinical populations may need to answer questions of clients concerning the use of CS, so a thorough understanding of the biology of Cr and the particular clinical condition is critical, as is staying current regarding the outcome of trials of CS. To date, human trials of CS in acute neurological injury such as TBI, stroke, and spinal cord injury are lacking. However, there is at least some evidence that CS may facilitate the rehabilitation process in spinal cord injury. In addition, although there are no data in this regard, it seems reasonable to hypothesize that CS might decrease the damage from brain injuries that might occur during sporting events. In contrast, human data in neurodegenerative clinical conditions is not particularly encouraging. Nonetheless, the lack of serious side effects reported to date suggests that there is no reason to specifically discourage CS by patients with neurodegenerative diseases, unless prohibited by their physician (e.g., owing to renal disease).

Modestly encouraging results are available in the literature for CS in muscle and metabolic diseases. At best though, CS will provide symptomatic benefits (e.g., McArdle's disease) and may delay disease progression, but CS is neither a panacea nor a cure. The most encouraging results appear to be in situations of combining CS with rehabilitative exercise (e.g., heart failure, COPD, and aging) where the primary target is skeletal muscle as opposed to neurological tissue, and CS is an adjunct to the exercise program. It is recommended that physician approval be granted before patients' implementing a CS program. Finally, based on the recommendations of Shao and Hathcock *(133)*, clients should be encouraged to keep the daily Cr dosage to ≤5 g/d.

REFERENCES

1. Wilken B, Ramirez JM, Probst I, Richter DW, Hanefeld F. Anoxic ATP depletion in neonatal mice brainstem is prevented by creatine supplementation. Arch Dis Child Fetal Neonatal Ed 2000; 82:F224–F227.
2. Greenhaff PL. The creatine-phosphocreatine system: there's more than one song in its repertoire. J Physiol 2001; 537:657.
3. Wyss M, Kaddurah-Daouk R. Creatine and creatinine metabolism. Physiol Rev 2000; 80:1107–1213.

4. Berger R, Middelanis J, Vaihinger HM, Mies G, Wilken B, Jensen A. Creatine protects the immature brain from hypoxic-ischemic injury. J Soc Gynecol Investig 2004; 11:9–15.
5. Hespel P, Op't Eijnde B, Van Leemputte M, et al. Oral creatine supplementation facilitates the rehabilitation of disuse atrophy and alters the expression of muscle myogenic factors in humans. J Physiol 2001; 536:625–633.
6. Beal MF. Mitochondria take center stage in aging and neurodegeneration. Ann Neurol 2005; 58:495–505.
7. Merenda A, Bullock R. Clinical treatments for mitochondrial dysfunctions after brain injury. Curr Opin Crit Care 2006; 12:90–96.
8. Raha S, Robinson BH. Mitochondria, oxygen free radicals, disease and ageing. Trends Biochem Sci 2000; 25:502–508.
9. Schlattner U, Tokarska-Schlattner M, Wallimann T. Mitochondrial creatine kinase in human health and disease. Biochim Biophys Acta 2006; 1762:164–180.
10. Hervias I, Beal MF, Manfredi G. Mitochondrial dysfunction and amyotrophic lateral sclerosis. Muscle Nerve 2006; 33:598–608.
11. Lawler JM, Barnes WS, Wu G, Song W, Demaree S. Direct antioxidant properties of creatine. Biochem Biophys Res Commun 2002; 290:47–52.
12. Sestili P, Martinelli C, Bravi G, et al. Creatine supplementation affords cytoprotection in oxidatively injured cultured mammalian cells via direct antioxidant activity. Free Radic Biol Med 2006; 40:837–849.
13. Bessman SP. The physiological significance of the creatine phosphate shuttle. Adv Exp Med Biol 1986; 194:1–11.
14. Saks VA, Kongas O, Vendelin M, Kay L. Role of the creatine/phosphocreatine system in the regulation of mitochondrial respiration. Acta Physiol Scand 2000; 168:635–641.
15. Smith SA, Montain SJ, Zientara GP, Fielding RA. Use of phosphocreatine kinetics to determine the influence of creatine on muscle mitochondrial respiration: an in vivo 31P-MRS study of oral creatine ingestion. J Appl Physiol 2004; 96:2288–2292.
16. Persky AM, Brazeau GA. Clinical pharmacology of the dietary supplement creatine monohydrate. Pharmacol Rev 2001; 53:161–176.
17. Friedlander, RM. Apoptosis and caspases in neurodegenerative diseases. N Engl J Med 2003; 348:1365–1375.
18. Salinska E, Danysz W, Lazarewicz JW. The role of excitotoxicity in neurodegeneration. Folia Neuropathol 2005; 43:322–339.
19. Armstrong JS. Mitochondrial membrane permeabilization: the sine qua non for cell death. Bioessays 2006; 28:253–260.
20. Parekh AB. Mitochondrial regulation of intracellular Ca2+ signaling: more than just simple Ca2+ buffers. News Physiol Sci 2003; 18:252–256.
21. Saris NE, Carafoli E. A historical review of cellular calcium handling, with emphasis on mitochondria. Biochemistry (Mosc) 2005; 70:187–194.
22. Crompton M. The mitochondrial permeability transition pore and its role in cell death. Biochem J 1999; 341(Pt 2):233–249.
23. O'Gorman E, Beutner G, Dolder M, Koretsky AP, Brdiczka D, Wallimann, T. The role of creatine kinase in inhibition of mitochondrial permeability transition. FEBS Lett 1997; 414:253–257.
24. Klivenyi P, Calingasan NY, Starkov A, et al. Neuroprotective mechanisms of creatine occur in the absence of mitochondrial creatine kinase. Neurobiol Dis 2004; 15:610–617.

25. Brustovetsky N, Brustovetsky T, Dubinsky JM. On the mechanisms of neuroprotection by creatine and phosphocreatine. J Neurochem 2001; 76:425–434.
26. Balestrino M, Lensman M, Parodi M, et al. Role of creatine and phosphocreatine in neuronal protection from anoxic and ischemic damage. Amino Acids 2002; 23:221–229.
27. Atlante A, Calissano P, Bobba A, Giannattasio S, Marra E, Passarella S. Glutamate neurotoxicity, oxidative stress and mitochondria. FEBS Lett 2001; 497:1–5.
28. Sonnewald U, Qu H, Aschner M. Pharmacology and toxicology of astrocyte-neuron glutamate transport and cycling. J Pharmacol Exp Ther 2002; 301:1–6.
29. Xu CJ, Klunk WE, Kanfer JN, Xiong Q, Miller G, Pettegrew JW. Phosphocreatine-dependent glutamate uptake by synaptic vesicles. A comparison with atp-dependent glutamate uptake. J Biol Chem 1996; 271:13,435–13,440.
30. Juravleva E, Barbakadze T, Mikeladze D, Kekelidze T. Creatine enhances survival of glutamate-treated neuronal/glial cells, modulates Ras/NF-kappaB signaling, and increases the generation of reactive oxygen species. J Neurosci Res 2005; 79:224–230.
31. Brewer GJ, Wallimann TW. Protective effect of the energy precursor creatine against toxicity of glutamate and beta-amyloid in rat hippocampal neurons. J Neurochem 2000; 74:1968–1978.
32. Malcon C, Kaddurah-Daouk R, Beal MF. Neuroprotective effects of creatine administration against NMDA and malonate toxicity. Brain Res 2000; 860:195–198.
33. Sullivan PG, Geiger JD, Mattson MP, Scheff SW. Dietary supplement creatine protects against traumatic brain injury. Ann Neurol 2000; 48:723–729.
34. Pena-Altamira E, Crochemore C, Virgili M, Contestabile A. Neurochemical correlates of differential neuroprotection by long-term dietary creatine supplementation. Brain Res 2005; 1058:183–188.
35. Wilkinson ID, Mitchel N, Breivik S, et al. Effects of creatine supplementation on cerebral white matter in competitive sportsmen. Clin J Sport Med 2006; 16:63–67.
36. Dechent P, Pouwels PJ, Wilken B, Hanefeld F, Frahm J. Increase of total creatine in human brain after oral supplementation of creatine-monohydrate. Am J Physiol 1999; 277:R698–R704.
37. Lyoo IK, Kong SW, Sung SM, et al. Multinuclear magnetic resonance spectroscopy of high-energy phosphate metabolites in human brain following oral supplementation of creatine-monohydrate. Psychiatry Res 2003; 123:87–100.
38. Leuzzi V, Bianchi MC, Tosetti M, et al. Brain creatine depletion: guanidinoacetate methyltransferase deficiency (improving with creatine supplementation). Neurology 2000; 55:1407–1409.
39. Stockler S, Holzbach U, Hanefeld F, et al. Creatine deficiency in the brain: a new, treatable inborn error of metabolism. Pediatr Res 1994; 36:409–413.
40. Stockler S, Hanefeld F, Frahm J. Creatine replacement therapy in guanidinoacetate methyltransferase deficiency, a novel inborn error of metabolism. Lancet 1996; 348:789, 790.
41. Bianchi MC, Tosetti M, Fornai F, et al. Reversible brain creatine deficiency in two sisters with normal blood creatine level. Ann Neurol 2000; 47:511–513.
42. Hersch SM, Gevorkian S, Marder K, et al. Creatine in Huntington disease is safe, tolerable, bioavailable in brain and reduces serum 8OH2'dG. Neurology 2006; 66:250–252.
43. Langlois J, Rutland-Brown W, Thomas K. Traumatic Brain Injury in the United States: Emergency Department Visits, Hospitalizations, and Deaths (Atlanta GA: US

Dept Health and Human Services, C. f. D. C. a. P., National Center for Injury Prevention and Control, ed.). 2006, pp. 1–55.
44. Thurman DJ, Branche CM, Sniezek JE. The epidemiology of sports-related traumatic brain injuries in the United States: recent developments. J Head Trauma Rehabil 1998; 13:1–8.
45. Scheff SW, Dhillon HS. Creatine-enhanced diet alters levels of lactate and free fatty acids after experimental brain injury. Neurochem Res 2004; 29:469–479.
46. Zhu S, Li M, Figueroa BE, et al. Prophylactic creatine administration mediates neuroprotection in cerebral ischemia in mice. J Neurosci 2004; 24:5909–5912.
47. Prass K, Royl G, Lindauer U, et al. Improved reperfusion and neuroprotection by creatine in a mouse model of stroke. J Cereb Blood Flow Metab 2006; 47:452–459.
48. Jackson CE, Bryan WW. Amyotrophic lateral sclerosis. Semin Neurol 1998; 18:27–39.
49. Strong MJ. The basic aspects of therapeutics in amyotrophic lateral sclerosis. Pharmacol Ther 2003; 98:379–414.
50. Klivenyi P, Ferrante RJ, Matthews RT, et al. Neuroprotective effects of creatine in a transgenic animal model of amyotrophic lateral sclerosis. Nat Med 1999; 5:347–350.
51. Andreassen OA, Jenkins BG, Dedeoglu A, et al. Increases in cortical glutamate concentrations in transgenic amyotrophic lateral sclerosis mice are attenuated by creatine supplementation. J Neurochem 2001; 77:383–390.
52. Snow RJ, Turnbull J, da Silva S, Jiang F, Tarnopolsky MA. Creatine supplementation and riluzole treatment provide similar beneficial effects in copper, zinc superoxide dismutase (G93A) transgenic mice. Neuroscience 2003; 119:661–667.
53. Zhang W, Narayanan M, Friedlander RM. Additive neuroprotective effects of minocycline with creatine in a mouse model of ALS. Ann Neurol 2003; 53:267–270.
54. Derave W, Van Den Bosch L, Lemmens G, Eijnde BO, Robberecht W, Hespel P. Skeletal muscle properties in a transgenic mouse model for amyotrophic lateral sclerosis: effects of creatine treatment. Neurobiol Dis 2003; 13:264–272.
55. Shefner JM, Cudkowicz ME, Schoenfeld D, et al. A clinical trial of creatine in ALS. Neurology 2004; 63:1656–1661.
56. Groeneveld GJ, Veldink JH, van der Tweel I, et al. A randomized sequential trial of creatine in amyotrophic lateral sclerosis. Ann Neurol 2003; 53:437–445.
57. Ryu H, Rosas HD, Hersch SM, Ferrante RJ. The therapeutic role of creatine in Huntington's disease. Pharmacol Ther 2005; 108:193–207.
58. Mazzini L, Balzarini C, Colombo R, et al. Effects of creatine supplementation on exercise performance and muscular strength in amyotrophic lateral sclerosis: preliminary results. J Neurol Sci 2001; 191:139–144.
59. Hausmann ON, Fouad K, Wallimann T, Schwab ME. Protective effects of oral creatine supplementation on spinal cord injury in rats. Spinal Cord 2002; 40:449–456.
60. Rabchevsky AG, Sullivan PG, Fugaccia I, Scheff SW. Creatine diet supplement for spinal cord injury: influences on functional recovery and tissue sparing in rats. J Neurotrauma 2003; 20:659–669.
61. Jacobs PL, Mahoney ET, Cohn KA, Sheradsky LF, Green BA. Oral creatine supplementation enhances upper extremity work capacity in persons with cervical-level spinal cord injury. Arch Phys Med Rehabil 2002; 83:19–23.
62. Kendall RW, Jacquemin G, Frost R, Burns SP. Creatine supplementation for weak muscles in persons with chronic tetraplegia: a randomized double-blind placebo-controlled crossover trial. J Spinal Cord Med 2005; 28:208–213.

63. NINDS_NET_PD_Investigators. A randomized, double-blind, futility clinical trial of creatine and minocycline in early Parkinson disease. Neurology 2006; 66:664–671.
64. Klivenyi P, Gardian G, Calingasan NY, Yang L, Beal MF. Additive neuroprotective effects of creatine and a cyclooxygenase 2 inhibitor against dopamine depletion in the 1-methyl-4-phenyl-1,2,3,6-tetrahydropyridine (MPTP) mouse model of Parkinson's disease. J Mol Neurosci 2003; 21:191–198.
65. Matthews RT, Ferrante RJ, Klivenyi P, et al. Creatine and cyclocreatine attenuate MPTP neurotoxicity. Exp Neurol 1999; 157:142–149.
66. Dedeoglu A, Kubilus JK, Yang L, et al. Creatine therapy provides neuroprotection after onset of clinical symptom in Huntington's disease transgenic mice. J Neurochem 2003; 85:1359–1367.
67. Jenkins BG, Koroshetz WJ, Beal MF, Rosen BR. Evidence for impairment of energy metabolism in vivo in Huntington's disease using localized 1H NMR spectroscopy. Neurology 1993; 43:2689–2695.
68. Koroshetz WJ, Jenkins BG, Rosen BR, Beal MF. Energy metabolism defects in Huntington's disease and effects of coenzyme Q10. Ann Neurol 1997; 41:160–165.
69. Sanchez-Pernaute R, Garcia-Segura JM, del Barrio Alba A, Viano J, de Yebenes JG. Clinical correlation of striatal 1H MRS changes in Huntington's disease. Neurology 1999; 53:806–812.
70. Matthews RT, Yang L, Jenkins BG, et al. Neuroprotective effects of creatine and cyclocreatine in animal models of Huntington's disease. J Neurosci 1998; 18:156–163.
71. Shear DA, Haik KL, Dunbar GL. Creatine reduces 3-nitropropionic-acid-induced cognitive and motor abnormalities in rats. Neuroreport 2000; 11:1833–1837.
72. Vis JC, de Boer-Van Huizen RT, Verbeek MM, de Waal RM, ten Donkelaar HJ, Kremer B. Creatine protects against 3-nitropropionic acid-induced cell death in murine corticostriatal slice cultures. Brain Res 2004; 1024:16–24.
73. Ferrante RJ, Andreassen OA, Jenkins BG, et al. Neuroprotective effects of creatine in a transgenic mouse model of Huntington's disease. J Neurosci 2000; 20:4389–4397.
74. Verbessem P, Lemiere J, Eijnde BO, et al. Creatine supplementation in Huntington's disease: a placebo-controlled pilot trial. Neurology 2003; 61:925–930.
75. Tabrizi SJ, Blamire AM, Manners DN, et al. Creatine therapy for Huntington's disease: clinical and MRS findings in a 1-year pilot study. Neurology 2003; 61:141–142.
76. Tabrizi SJ, Blamire AM, Manners DN, et al. High-dose creatine therapy for Huntington disease: a 2-year clinical and MRS study. Neurology 2005; 64:1655, 1656.
77. Bender A, Auer DP, Merl T, et al. Creatine supplementation lowers brain glutamate levels in Huntington's disease. J Neurol 2005; 252:36–41.
78. Kent-Braun JA, Sharma KR, Miller RG, Weiner MW. Postexercise phosphocreatine resynthesis is slowed in multiple sclerosis. Muscle Nerve 1994; 17:835–841.
79. Lambert CP, Archer RL, Carrithers JA, Fink WJ, Evans WJ, Trappe TA. Influence of creatine monohydrate ingestion on muscle metabolites and intense exercise capacity in individuals with multiple sclerosis. Arch Phys Med Rehabil 2003; 84:1206–1210.

80. Stout JR, Eckerson JM, May E, Coulter C, Bradley-Popovich GE. Effects of resistance exercise and creatine supplementation on myasthenia gravis: a case study. Med Sci Sports Exerc 2001; 33:869–872.
81. Tarnopolsky MA, Raha S. Mitochondrial myopathies: diagnosis, exercise intolerance, and treatment options. Med Sci Sports Exerc 2005; 37:2086–2093.
82. Tarnopolsky MA. What can metabolic myopathies teach us about exercise physiology? Appl Physiol Nutr Metab 2006; 31:21–30.
83. Tarnopolsky MA, Simon DK, Roy BD, et al. Attenuation of free radical production and paracrystalline inclusions by creatine supplementation in a patient with a novel cytochrome b mutation. Muscle Nerve 2004; 29:537–547.
84. Tarnopolsky MA, Roy BD, MacDonald JR. A randomized, controlled trial of creatine monohydrate in patients with mitochondrial cytopathies. Muscle Nerve 1997; 20:1502–1509.
85. Borchert A, Wilichowski E, Hanefeld F. Supplementation with creatine monohydrate in children with mitochondrial encephalomyopathies. Muscle Nerve 1999; 22:1299–300.
86. Klopstock T, Querner V, Schmidt F, et al. A placebo-controlled crossover trial of creatine in mitochondrial diseases. Neurology 2000; 55:1748–1751.
87. Pulido SM, Passaquin AC, Leijendekker WJ, Challet C, Wallimann T, Ruegg UT. Creatine supplementation improves intracellular Ca2+ handling and survival in mdx skeletal muscle cells. FEBS Lett 1998; 439:357–362.
88. Felber S, Skladal D, Wyss M, Kremser C, Koller A, Sperl W. Oral creatine supplementation in Duchenne muscular dystrophy: a clinical and 31P magnetic resonance spectroscopy study. Neurol Res 2000; 22:145–150.
89. Louis M, Raymackers JM, Debaix H, Lebacq J, Francaux M. Effect of creatine supplementation on skeletal muscle of mdx mice. Muscle Nerve 2004; 29:687–692.
90. Passaquin AC, Renard M, Kay L, Challet C, Mokhtarian A, Wallimann T, Ruegg UT. Creatine supplementation reduces skeletal muscle degeneration and enhances mitochondrial function in mdx mice. Neuromuscul Disord 2002; 12:174–182.
91. Walter MC, Lochmuller H, Reilich P, et al. Creatine monohydrate in muscular dystrophies: A double-blind, placebo-controlled clinical study. Neurology 2000; 54:1848–1850.
92. Louis M, Lebacq J, Poortmans JR, et al. Beneficial effects of creatine supplementation in dystrophic patients. Muscle Nerve 2003; 27:604–610.
93. Tarnopolsky MA, Mahoney DJ, Vajsar J, et al. Creatine monohydrate enhances strength and body composition in Duchenne muscular dystrophy. Neurology 2004; 62:1771–1777.
94. Escolar DM, Buyse G, Henricson E, et al. CINRG randomized controlled trial of creatine and glutamine in Duchenne muscular dystrophy. Ann Neurol 2005; 58:151–155.
95. Ranum LP, Cooper TA. Rna-Mediated Neuromuscular Disorders. Annu Rev Neurosci 2006; 29:259–277.
96. Tramonte JJ, Burns TM. Myotonic dystrophy. Arch Neurol 2005; 62:1316–1319.
97. Tarnopolsky M, Mahoney D, Thompson T, Naylor H, Doherty TJ. Creatine monohydrate supplementation does not increase muscle strength, lean body mass, or muscle phosphocreatine in patients with myotonic dystrophy type 1. Muscle Nerve 2004; 29:51–58.

98. Walter MC, Reilich P, Lochmuller H, et al. Creatine monohydrate in myotonic dystrophy: a double-blind, placebo-controlled clinical study. J Neurol 2002; 249:1717–1722.
99. Biggar WD, Klamut HJ, Demacio PC, Stevens DJ, Ray PN. Duchenne muscular dystrophy: current knowledge, treatment, and future prospects. Clin Orthop Relat Res 2002; 401:88–106.
100. Haller RG. Treatment of McArdle disease. Arch Neurol 2000; 57:923, 924.
101. Vorgerd M, Grehl T, Jager M, et al. Creatine therapy in myophosphorylase deficiency (McArdle disease): a placebo-controlled crossover trial. Arch Neurol 2000; 57:956–963.
102. Zange J, Kornblum C, Muller K, et al. Creatine supplementation results in elevated phosphocreatine/adenosine triphosphate (ATP) ratios in the calf muscle of athletes but not in patients with myopathies. Ann Neurol 2002; 52:126; author reply 126, 127.
103. Vorgerd M, Zange J, Kley R, et al. Effect of high-dose creatine therapy on symptoms of exercise intolerance in McArdle disease: double-blind, placebo-controlled crossover study. Arch Neurol 2002; 59:97–101.
104. Sipila I, Rapola J, Simell O, Vannas A. Supplementary creatine as a treatment for gyrate atrophy of the choroid and retina. N Engl J Med 1981; 304:867–870.
105. Heinanen K, Nanto-Salonen K, Komu M, et al. Creatine corrects muscle 31P spectrum in gyrate atrophy with hyperornithinaemia. Eur J Clin Invest 1999; 29:1060–1065.
106. Nanto-Salonen K, Komu M, Lundbom N, et al. Reduced brain creatine in gyrate atrophy of the choroid and retina with hyperornithinemia. Neurology 1999; 53:303–307.
107. Vannas-Sulonen K, Sipila I, Vannas A, Simell O, Rapola J. Gyrate atrophy of the choroid and retina. A five-year follow-up of creatine supplementation. Ophthalmology 1985; 92:1719–1727.
108. Pina IL, Apstein CS, Balady GJ, et al. Exercise and heart failure: A statement from the American Heart Association Committee on exercise, rehabilitation, and prevention. Circulation 2003; 107:1210–1225.
109. Neubauer S, Remkes H, Spindler M, et al. Downregulation of the Na(+)-creatine cotransporter in failing human myocardium and in experimental heart failure. Circulation 1999; 100:1847–1850.
110. Nakae I, Mitsunami K, Matsuo S, et al. Myocardial creatine concentration in various nonischemic heart diseases assessed by 1H magnetic resonance spectroscopy. Circ J 2005; 69:711–716.
111. Nakae I, Mitsunami K, Matsuo S, et al. Assessment of myocardial creatine concentration in dysfunctional human heart by proton magnetic resonance spectroscopy. Magn Reson Med Sci 2004; 3:19–25.
112. Weiss RG, Gerstenblith G, Bottomley PA. ATP flux through creatine kinase in the normal, stressed, and failing human heart. Proc Natl Acad Sci USA 2005; 102:808–813.
113. Nakae I, Mitsunami K, Omura T, et al. Proton magnetic resonance spectroscopy can detect creatine depletion associated with the progression of heart failure in cardiomyopathy. J Am Coll Cardiol 2003; 42:1587–1593.
114. Gordon A, Hultman E, Kaijser L, et al. Creatine supplementation in chronic heart failure increases skeletal muscle creatine phosphate and muscle performance. Cardiovasc Res 1995; 30:413–418.

115. Andrews R, Greenhaff P, Curtis S, Perry A, Cowley AJ. The effect of dietary creatine supplementation on skeletal muscle metabolism in congestive heart failure. Eur Heart J 1998; 19:617–622.
116. Duscha BD, Annex BH, Green HJ, Pippen AM, Kraus WE. Deconditioning fails to explain peripheral skeletal muscle alterations in men with chronic heart failure. J Am Coll Cardiol 2002; 39:1170–1174.
117. Kuethe F, Krack A, Richartz BM, Figulla HR. Creatine supplementation improves muscle strength in patients with congestive heart failure. Pharmazie 2006; 61:218–222.
118. Sala E, Roca J, Marrades RM, et al. Effects of endurance training on skeletal muscle bioenergetics in chronic obstructive pulmonary disease. Am J Respir Crit Care Med 1999; 159:1726–1734.
119. Fuld JP, Kilduff LP, Neder JA, et al. Creatine supplementation during pulmonary rehabilitation in chronic obstructive pulmonary disease. Thorax 2005; 60:531–537.
120. Rawson ES, Volek JS. Effects of creatine supplementation and resistance training on muscle strength and weightlifting performance. J Strength Cond Res 2003; 17:822–831.
121. Tyler TF, Nicholas SJ, Hershman EB, Glace BW, Mullaney MJ, McHugh MP. The effect of creatine supplementation on strength recovery after anterior cruciate ligament (ACL) reconstruction: a randomized, placebo-controlled, double-blind trial. Am J Sports Med 2004; 32:383–388.
122. Willoughby DS, Rosene JM. Effects of oral creatine and resistance training on myogenic regulatory factor expression. Med Sci Sports Exerc 2003; 35:923–929.
123. Rawson ES, Wehnert ML, Clarkson PM. Effects of 30 days of creatine ingestion in older men. Eur J Appl Physiol Occup Physiol 1999; 80:139–144.
124. Rawson ES, Clarkson PM, Price TB, Miles MP. Differential response of muscle phosphocreatine to creatine supplementation in young and old subjects. Acta Physiol Scand 2002; 174:57–65.
125. Rawson ES, Clarkson PM. Acute creatine supplementation in older men. Int J Sports Med 2000; 21:71–75.
126. Gotshalk LA, Volek JS, Staron RS, Denegar CR, Hagerman FC, Kraemer WJ. Creatine supplementation improves muscular performance in older men. Med Sci Sports Exerc 2002; 34:537–543.
127. Brose A, Parise G, Tarnopolsky MA. Creatine supplementation enhances isometric strength and body composition improvements following strength exercise training in older adults. J Gerontol A Biol Sci Med Sci 2003; 58:11–19.
128. Chrusch MJ, Chilibeck PD, Chad KE, Davison KS, Burke DG. Creatine supplementation combined with resistance training in older men. Med Sci Sports Exerc 2001; 33:2111–2117.
129. Eijnde BO, Van Leemputte M, Goris M, et al. Effects of creatine supplementation and exercise training on fitness in men 55–75 yr old. J Appl Physiol 2003; 95:818–828.
130. Bermon S, Venembre P, Sachet C, Valour S, Dolisi C. Effects of creatine monohydrate ingestion in sedentary and weight-trained older adults. Acta Physiol Scand 1998; 164:147–155.
131. Chilibeck PD, Chrusch MJ, Chad KE, Shawn Davison K, Burke DG. Creatine monohydrate and resistance training increase bone mineral content and density in older men. J Nutr Health Aging 1998; 9:352, 353.

132. Burke DG, Chilibeck PD, Parise G, Candow DG, Mahoney D, Tarnopolsky M. Effect of creatine and weight training on muscle creatine and performance in vegetarians. Med Sci Sports Exerc 2003; 35:1946–1955.
133. Shao A, Hathcock JN. Risk assessment for creatine monohydrate. Regul Toxicol Pharmacol 2006; 45:242–251.

7 Creatine Overview
Facts, Fallacies, and Future

Mike Greenwood, PhD

1. INTRODUCTION

Without question, since its over the counter availability to consumers in 1992, creatine has become one of the most popular nutritional supplements among exercise and sport populations. In addition to its popularity, creatine has become one of the most extensively studied and research validated products that have been experimentally dissected in a multitude of ways. Specifically, investigators have evaluated topics such as muscle-creatine content and phosphocreatine resynthesis, short- and long-term ergogenic effects of creatine ingestion, gender issues associated with creatine ingestion, age-specific issues related to creatine ingestion, ethical considerations of creatine ingestion, viable clinical and medical applications of creatine ingestion, health and safety concerns regarding creatine ingestion, and more recently relevant biochemical mechanisms regarding the creatine transport system. Although each of these research approaches have greatly contributed to the body of creatine literature, it is first imperative to grasp various foundational aspects associated with understanding this controversial nutritional supplement. With these considerations in mind, the purpose of this chapter is to set the stage for a creatine overview regarding the following information: (1) creatine facts, fallacies, and safety (2) creatine quality, purity, and formulations, (3) creatine dosage protocols, (4) creatine nutritional supplement combinations, (5) foundational creatine ergogenic efficacy, (6) future creatine research options, and (7) common creatine practical applications.

From: *Essentials of Creatine in Sports and Health*
Edited by: J. R. Stout, J. Antonio and D. Kalman © Humana Press Inc., Totowa, NJ

2. REVIEW OF LITERATURE

2.1. Creatine Facts, Fallacies, and Safety: Anecdotol Misconceptions Vs Science

Although creatine has been closely evaluated for over a decade (>1000 published studies at the time of this writing), many questions and concerns continue to surface primarily revolving around the issue of safety. Although the current valid and quality science surrounding creatine supplementation speaks for itself, it is amazing to this author the amount of misinformation that still continues to surface on the topic. Unfortunately, along with the quality of creatine research that does exist, this popular nutritional aid has been anecdotally dissected and misinterpreted. One major cause of this anecdotal misinformation plague has been popular media venues (i.e., television, radio, newspapers, and lay journals) disseminating information that is not educationally and/or scientifically based. In a perfect world, this author wishes that individuals consumed with this topic would perform their homework through the plethora of science-based reading material and/or contacting knowledgeable professionals to attain accurate information on the subject. Before diving head first into a variety of common anecdotal misconceptions that exist on this topic, the readership must be informed and fully understand that the only clinically significant reported side effect associated with creatine ingestion to date has been weight gain *(1–3)*.

Numerous anecdotal side effects have been reported in the popular literature and media venues that have created a variety of misconceptions and confusion surrounding creatine supplementation. Some of the more ridiculous fallacies that have surfaced include creatine ingestion causing the following: lower limb amputations, kidney stones, cancer, soft soles of the feet, dry skeletal muscle, and my favorite myth, creatine the anabolic steroid. Although there is no science behind these specific anecdotal claims, a host of other more rationale physiological safety questions have been raised and denounced as problematic issues associated with creatine ingestion. For instance, concerns have surfaced over the medical/health safety issues of creatine supplementation regarding select populations even though extensive research has not divulged any negative side effects in the areas of endogenous creatine synthesis *(4,5)*, renal and liver function *(3,6–14)*, muscle and liver enzyme efflux *(3,6,8)*, blood volume *(15–19)*, electrolyte status *(3,19,20)*, blood pressure *(21,22)*, and/or general markers of medical safety *(6,8,12,16,18,22–24)*. It should also be noted that select researchers have claimed that creatine ingestion increases the risk of anterior compartment pressure in the lower limbs thereby increasing the probability of developing anterior compartment syndrome complications *(25–28)*. For over a decade, numerous scientifically based double-blind placebo-controlled studies have assessed the medical safety of creatine supplementation and none of these studies have supported the previously mentioned anecdotal problems or concerns associated with the likelihood of developing anterior compartment syndrome complications *(6,8,12–16,29–32)*. In fact, investigators have revealed time and time again that the incidence of negative safety and injury occurrences in creatine users does not appear to be greater than subjects who ingest placebos and in select cases have actually reported fewer complications *(16,29,30,33)*.

Specifically, there have been a variety of anecdotal reports that creatine supplementation during intense training in hot/humid environments may predispose athletes to increase incidence of muscle cramping, dehydration, and/or musculoskeletal injuries such as muscle strains *(17,28,34)*. Fortunately, these unsubstantiated anecdotal concerns revolving around creatine supplementation and athletic injuries have generated time and labor intensive field research in this area. Specifically, researchers evaluated the effects of creatine supplementation on the incidence of injury observed during 3-yr of National Collegiate Athletic Association (NCAA) division IA college football training and competition *(33)*. One-hundred athletes participating in the 1998–2000 football seasons ingested creatine following workouts and practices (15.75 g of creatine for 5 d followed by ingesting an average of 5 g/d, thereafter administered in 5–10 g doses). These athletes practiced or played in environmental conditions ranging from 8 to 40°C (mean 24.7 ± 5°C) and 19–98% relative humidity (49.3 ± 17%). It was found that creatine users had fewer incidence of cramping (37 incidence by creatine users to 96 incidence by nonusers, 39%), heat/dehydration (8:28, 36%), muscle tightness (18:42, 43%), muscle pulls/strains (25:51, 49%), noncontact joint injuries (44:32, 33%), contact injuries (39:104, 44%), illness (12:27, 44%), number of missed practices owing to injury (19:41, 46%), players lost for the season (3:8, 38%), and total injuries/missed practices (205:529, 39%). It was concluded that creatine supplementation does not increase the incidence of injury or cramping in division IA college football players. In conjunction with this study, investigators also evaluated the long-term safety of creatine supplementation over a 21-mo time frame *(16)*. The results indicated that 5 g/d administered in 5–10 g/d doses does not adversely affect health status markers in athletes undergoing intense training compared with athletes who were not taking creatine. This health status evaluation included fasting blood and 24-h urine samples (Table 1).

In a similar study investigators examined the effects of creatine supplementation on cramping and injury occurrence in 38 division IA football players over a single competitive season *(29)*. These athletes

Table 1
Observed Injury Rates For Creatine Vs Noncreatine Users: Division IA Football Team ($n = 98$)

Treated injuries n = 54/44	*Number of injuries (users/nonusers)*	*Percentage-creatine users with injuries (%)*
Cramping	37/96	39
Heat/dehydration	8/28	36
Muscle tightness	18/42	43
Muscle pulls/strains	25/51	49
Noncontact injuries	44/132	33
Contact injuries	39/104	44
Illness	12/27	44
Missed practices	19/41	46
Players lost season	3/8	38
Total injuries	205/529	39

trained, practiced, or played in environmental conditions ranging from 15 to 37°C (mean 27.26 ± 10.93°C) and 46–91% relative humidity (54.17 ± 9.71%) and ingested 0.3 g/kg/d of creatine for 5 d followed by consuming an average of 0.03 g/kg/d thereafter following workouts, practices, and games. The findings in this study revealed that the incidence of cramping, dehydration, muscle tightness, muscle strains, and total injuries among creatine users were significantly less than the nonusers. Moreover, the incidence of noncontact injuries, contact injuries, illness, missed practices, and players lost with a season ending injury observed in the creatine users were similar or lower than the nonusers. Once again, these findings support that creatine supplementation does not increase the incidence of injury or cramping in division IA college football players (Table 2).

Finally, in a recent investigation, researchers examined the effects of creatine supplementation on cramping and injury of 21 division I collegiate baseball players during 6 mo of training and/or competition *(30)*. These athletes ingested 15–25 g/d of creatine for 5 d followed by 5 g/d of creatine after practices or games. Environmental conditions for these athletes during training and competition ranged from 27 to 35°C (mean 30.4 ± 0.6°C) and 59–91% relative humidity (77.1 ± 2.53%). Although no heat/dehydration events were reported for either group, the creatine-users had significantly fewer total injuries than the noncreatine users. In addition, there were no significant differences between groups regarding cramping, muscle tightness, muscle strains, noncontact joint injuries, contact injuries, illness, missed practices owing to injury, and players

Table 2
Observed Injury Rates For Creatine Users Vs Noncreatine Users: Division IA Football Team ($n = 72$)

Treated injuries n = 38/34	*Number of injuries (users/nonusers)*	*Percentage-creatine users with injuries (%)*
Cramping	19/47	40
Heat/dehydration	4/13	31
Muscle tightness	9/29	39
Muscle pulls/strains	28/62	45
Noncontact injuries	22/60	37
Contact injuries	89/176	51
Illness	29/69	42
Missed practices	52/138	38
Players lost season	1/2	50
Total injuries	253/595	43

lost for the season. Based on the findings in this investigation, creatine supplementation during collegiate baseball training and competition did not increase the incidence of injury or cramping. Although additional research is warranted to evaluate the effect of creatine supplementation on athletes training in hot/humid climates, research to date consistently indicates that creatine is medically safe and does not increase the incidence of cramping or injury (Table 3).

Therefore, the current research findings surrounding creatine supplementation and athletic injuries may help professionals involved in the training and/or medical supervision of athletes (i.e., athletic coaches, certified athletic trainers, researchers, certified strength and conditioning specialists, nutritional consultants, administrators, and athletic governing bodies) to better examine the methods used to train and/or manage athletes (i.e., consecutive two-a-day training bouts in extreme ambient climates, exhaustive conditioning drills, hydration strategies, and so on). In this regard, it appears that the type of activities and conditions that athletes are asked to train and/or compete in might place them at a greater risk of dehydration, cramping and/or injury than anecdotally associating these problems with creatine supplementation.

Finally, it is also imperative to address a commonly asked question regarding creatine supplementation. Specifically, are the long-term side effects of creatine ingestion safe? Before creatine was available to the general public in the 1990s, athletes across the world were utilizing creatine as an effective ergogenic aid even in the 1960s. Even though there have been no clinically significant negative side effects directly attributed to

Table 3
Observed Injury Rates For Creatine Users Vs Noncreatine Users: Division IA Baseball Players ($n = 39$)

Treated injuries *n = 21/18*	*Number of injuries (users/nonusers)*	*Percentage-creatine users with injuries (%)*
Cramping	4/8	50
Heat/dehydration	0/0	0
Muscle tightness	7/11	64
Muscle pulls/strains	6/10	60
Noncontact injuries	5/7	71
Contact injuries	4/8	50
Illness	8/8	1

creatine supplementation to date, some question the long-term safety outcomes of this popular product. Consequently, researchers approached this concern to help scientifically address this important medical query. At the time of this writing, no long-term negative side effects have been observed in athletic populations lasting up to 5 yr, infants with creatine synthesis deficiency lasting up to 3 yr, or in patient populations lasting up to 5 yr *(2,6,8,12,16)*. In two very valuable time and labor intensive studies, patients ingesting 1.5–3 g of creatine per day have been safely monitored over a 25-yr period (1981) reporting no significant side effects *(35,36)*. In addition, researchers have demonstrated a variety of potentially helpful clinical uses of creatine for: infants/patients limited with creatine synthesis deficiencies; patients suffering from various orthopedic injuries; patients with a variety of neuromuscular afflictions as well as patients with select heart conditions. Consequently, all available evidence suggests that long-term creatine supplementation appears to be safe and has been shown to be effective as a therapeutic aid with select clinical populations. In this author's opinion, the medical profession's use of creatine as a potential therapeutic aid has helped banish as well as educate many of the anticreatine critics but the educational process is dynamic.

2.2. Creatine Quality, Purity, and Formulation

One important common sense scientific investigative approach promoted in the authors' laboratory research is to determine the safety and efficacy of any nutritional supplement evaluated. It is the philosophical stand that the public has a right to know the truth regarding these two important concepts and any validated information attained after that is icing on the cake. Before starting any study it is considered imperative to test every product by determining if ingredient formulations and amounts noted on the dietary label are indeed accurate. It is impossible to make any valid inferential scientific claims of research outcomes without answering these factors first. The consumer and/or sport nutrition specialist should attempt to establish educationally informed strategies for evaluating dietary supplement labels, which often help develop key dietary supplement knowledge. Utilizing this approach will also assist the individual in making informed decisions regarding reputable sport supplement companies/products/sources. If general consumers do not have direct access to effectively determine these strategies, researchers from reputable sport nutrition laboratories and/or certified sport nutritionists can be valuable resources to effectively accomplish this process. At the time of this writing, the top two manufacturing sources of raw creatine monohydrate are the United States and Germany. Wide spread distribution of creatine monohydrate can also be found in China but various impurities (i.e., creatinine, dicyandiamide, and/or dihydrotriazine) have been detected by some investigators *(37,38)*. Hence, extreme caution should be used when buying creatine monohydrate in an attempt to locate the highest quality formulations possible. In addition, it is advisable to purchase ones' supplements from inspected facilities that produce and maintain good Food and Drug Administration (FDA) manufacturing practice standards *(2)*.

As specific safety issues surrounding creatine ingestion have been previously addressed, the quality, purity, and formulations of this popular supplement are next on the agenda. Specifically, the vast majority of studies conducted with creatine products have utilized powered formulations of pharmacological grade creatine monohydrate as well as more expensive formulations of oral or intravenous phosphocreatine primarily used in various medical circumstances. In an effort to make the ingestion of creatine products more accessible to target audiences, the supplement industry has developed a variety of product formulations of creatine to entice the consumer to purchase products based on variety, convenience, and in some instances individual palatability. Although a plethora of creatine supplements can be purchased in the forms of candy, gum, tablets, capsules, bars, effervescent, citrate, ethyl ester, and serum/liquid, there is no supporting data that reveals that these alternatives are any more effective than a quality processed creatine monohydrate. These specially marketed products are also more expensive than a quality creatine monohydrate. Often, this conglomerate of creatine formulations are touted as more effective for increasing exercise capacity as well as better creatine uptake to the muscle. Again, this is why anecdotal claims need to be replaced with quality research outcomes. For example, in a study conducted by Kreider and colleagues, a type of serum creatine was found to have no effect on muscle-creatine uptake even with dosages far above the suggested amount *(39)*. Furthermore, other published investigations have compared the efficacy of creatine candy, gum, and effervescent, with creatine monohydrate, as the ergogenic values of ingesting these products were no different *(19,39–42)*. It should also be noted that there is no scientific support that ingesting higher or lower dosage amounts of these unique creatine products will accomplish the same or better outcomes than consuming creatine monohydrate relative to lower degradation in the stomach, better intestinal absorption, faster absorption in the blood, and/or greater muscle uptake.

2.2.1. Creatine Dosage Protocols

For nearly 15 yr investigators have evaluated specific creatine dosage protocols in an effort to determine viable ingestion uptake methods. Although it has been established that the quality, purity, and formulation of the creatine monohydrate is a critical component in this process, the optimal dosage protocols during the loading and maintenance phases are also a critical consideration. The amount of creatine ingestion becomes critical as some professionals indicate nonsignificant research findings have been primarily attributed to inadequate ingestion protocols such as poor quality/purity, improper dosage amounts/durations of ingestion, low statistical power, no incorporation of experimental controls, and failure to implement testing/training of the phosphocreatine energy system into the research design *(1)*.

To date, the most reported method cited in the scientific literature to elevate muscle-creatine stores is by consuming 0.3 g/kg/d of creatine monohydrate for 5–7 d in the "loading phase," which for many individuals equates to 5 g ingested four times per day *(1,2)*. Although responses to creatine uptake levels are individualized, this creatine dosage method has been shown to effectively increase muscle creatine and phosphocreatine ranging from 10 to 40% *(1)*. Furthermore, once muscle-creatine stores have reached optimal saturation levels, ingesting 3–5 g (0.03 g/kg/d—maintenance phase) of creatine monohydrate daily can effectively help maintain elevated creatine stores *(1,2)*. Scientific findings such as these have led researchers to ponder and investigate whether different creatine dosage protocols might have similar significant scientific effects. For example, some researchers recommend that an effective supplementation protocol is to ingest 3 g/d of creatine monohydrate for 28-d thereby eliminating the common "loading phase" *(43)*. In a recent study by Willoughby and colleagues, ingesting 6 g/d of creatine over 12 wk of resistance training revealed improvements regarding strength and muscle mass *(44)*. Although this approach has been shown to be scientifically beneficial, it is evident more time is required for optimal creatine saturation levels to be attained. Thus, if one desires the benefits of creatine uptake more rapidly, utilizing the loading phase still has benefits, but if time is not of the essence, a lower maintenance dosage accomplishes similar results across longer time-periods.

The excitement and passion affiliated with conducting quality creatine research has led investigators to other valuable contributions regarding creative creatine ingestion strategies. Specifically, although the majority of creatine research has centered around creatine dosages based on total body weight, Burke and associates investigated creatine dosages related to lean body mass *(45)*. Compared with typical creatine ingestion strategies, Burke's research group promoted a loading dose of 0.25 g/kg lbm/d for 7 d and maintenance dosages of 0.0625 g/kg lbm/d to control creatine stores there after. In this study, fruit juice was also administered to enhance intramuscular concentrations of creatine and phosphocreatine accumulations. In utilizing this dosing approach, it should be noted that lean body mass only included soft tissue analysis thereby excluding bone mineral content as a viable dependent variable. Although additional research is certainly warranted with respect to this creatine dosing strategy, the author is at liberty to state that there are a series of creatine investigations scheduled in the Exercise Biochemistry and Nutrition Laboratory this year related to this and other creatine-related issues.

To conclude this section, it is important to present relevant information related to other commonly asked questions surrounding creatine supplementation. Specifically, what are the best nutritional ingestion timing methods to follow regarding creatine supplementation? As intense exercise increases anabolic hormone release, many professionals promote the ingestion of carbohydrates, proteins, and/or essential amino acids as soon as possible after the training bout in an effort to elevate glycogen resynthesis while also enhancing protein synthesis *(46–50)*. This scientific mechanism is attributed to increased insulin levels promoted by the ingestion of carbohydrate and protein with creatine thereby further igniting protein synthesis through essential amino acid mechanisms. Again, because insulin levels enhance creatine uptake, ingestion of creatine after exercise with a carbohydrate and/or protein supplement is considered a viable method to promote and/or maintain intracellular muscle-creatine levels and stores. Various creatine supplement companies promote this method of creatine with carbohydrate/protein ingestion 1–2 h before and after exercise regarding the rationale noted earlier with the understanding that this concept, like any other nutritional supplement, requires verifiable research outcomes behind it to warrant its effective infusion into the exercise/training regime. If an individual is concerned about body weight management outcomes (losing or gaining excess body weight), caution should be utilized when establishing creatine/carbohydrate/protein strategies as strength to mass ratios can be negatively altered. The topic of creatine ingestion combined with other nutritional supplements is addressed in detail within the next section of this chapter.

Another typical question that surfaces on a regular basis relates to creatine cycling strategies. Specifically, from a scientific perspective, should I follow cycling protocols when taking creatine supplementation? To date, there is no research to support that cycling on and off creatine supplementation is any more or less beneficial than following established loading/maintenance strategies. Various professionals promote cycling off creatine when training intensities are reduced to accommodate specific aspects of the periodization cycles. Often, this approach is based on an athletic or strength training coach's personal philosophy toward creatine supplementation. Thus, the prominent gains associated with this creatine supplementation approach would be enhancing various exercise/athletic populations to attain optimal activity and health specific potentials during heavy training cycles *(16,51)*.

At times inquiries arise related to the scientific rationale for ingesting creatine four different times during the loading phase. Although this issue is not a frequently asked question, it is unique to the creatine craze and has scientific merit behind it. Common sense dictates this approach, as the consumption of 20–25 g of any product would not sit well with the digestive system. In addition, for those unaccustomed to ingesting creatine, consuming an over abundance of this popular ergogenic may take additional adjustment time. This statement would hold true regarding a variety of nutritional supplements, which is another valid reason to follow the prescribed protocols noted on the label which is hopefully research based. Another common sense explanation for ingesting creatine four times a day at 4-h time intervals equates to better absorption time to allow for a viable uptake of creatine stores. Dispersing and ingesting this endogenous product according to this protocol helps regulate and accomplish this physiological process more effectively than consuming 20–25 g of creatine in a single dosage. Scientific evidence suggests, after ingesting 5 g of creatine monohydrate the plasma-creatine levels elevate up to 10 times the normal level within 1 h. This process increases the blood muscle concentration gradient thereby transporting and storing more blood borne intracellular muscle creatine *(52)*. Considering blood creatine has a relatively limited half-life of approx 1.5 h, it becomes imperative to follow a closely timed ingestion regimen to bolster this process *(53)*. Implementing this approach in the loading phase,

Table 4
Creatine Ingestion Strategies

Prompt loading and maintenance strategies
• 20/25 g/d divided into 4/5 equal doses (5–7 d)
• 0.3 g/kg/d (15–30 g/d for 50–100 kg individuals) (5–7 d)
• 0.25 g/kg lean body mass/d (5–7 d) 0.625 g/kg lean body mass/d after load/maintenance
• 3–5 g/d (0.03 g/kg/d) after load/maintenance training phase
Gradual low dose strategy
• 3–6 g/d maintenance phase no urgent load ingestion
Cycle on and off strategy
• Loading and maintenance dosage based on person's training intensity, body weight, or lbm while refraining during nontraining segments

considering one agrees with the loading method, will place emphasis on consistent plasma creatine levels concentrations, which will be effectively stored in the muscle throughout the course of the day. Once creatine stores are saturated, the individual may advance to the typical maintenance phase. Hypothetically, as elevated creatine stores return to baseline within 4–6 wk, a maintenance protocol of creatine (3–5 g) is effective in increasing and retaining elevated creatine stores over time especially when consumed with carbohydrate and/or protein combinations *(54)* (Table 4).

Finally, questions often surface regarding creatine supplementation as a potential and effective therapeutic option for various clinical populations. For example, creatine synthesis deficiencies, gyrate atrophy, neuromuscular diseases, various heart conditions, brain and spinal cord injuries, arthritis, diabetes, and high cholesterol and triglyceride levels *(1,2)*. In addition, emphasis is often placed on excessive amounts of creatine supplement related to ergogenic benefits. It should be noted that much higher creatine dosages have been prescribed in the medical profession to treat clinical patients suffering from various afflictions whereas no negative issues regarding safety in these instances have been shown *(2)*. Details surrounding the therapeutic values of creatine ingestion with various medical conditions are covered in another chapter of this textbook.

2.2.1.1. Creatine Nutritional Supplement Combinations. There has been tremendous interest in determining effective strategies to enhance muscle uptake of creatine. Combining various macronutrients, primarily carbohydrate and protein, with creatine, has been a major source of interest in the last few years in an effort to determine the most effective methods of enhancing transport systems that optimize creatine storage through greater intestinal and/or intracellular muscle uptake. Various combinations and timing mechanisms have been implemented to determine optimal combinations to promote creatine efficacy. Numerous studies have shown that adding macronutrients to creatine has promoted better muscular retention of creatine. Specifically, Greenwood and colleagues *(55)* reported that dextrose added to creatine resulted in better retention than creatine monohydrate alone, whereas other research teams reported that adding carbohydrate and protein to creatine is more effective for promoting muscle retention of creatine *(56)*. Green and associates also reported that adding carbohydrate to creatine supplementation increased total muscle creatine by 60% *(57)*.

The previously mentioned studies make an interesting contrast to the recent literature that exists concerning adding macronutrients to creatine and the improvement of strength and lean body mass. Burke and colleagues reported that whey protein plus creatine monohydrate increased lean body mass and bench press strength to a greater extent than whey protein alone, but no comparison was made with creatine monohydrate *(58)*. Tarnopolsky and associates reported that postexercise protein and carbohydrate supplementation was as effective as a creatine/carbohydrate combination in improving lean body mass and strength *(59)*. Researchers have also reported that a postexercise whey protein, amino acids, creatine, and carbohydrate was no more effective at promoting muscle strength and endurance, but a trend toward increasing lean body mass was seen *(60)*. In another study, no performance advantage was gained by adding carbohydrate to a creatine supplement over the course of 4 wk with competitive swimmers *(61)*. Finally, it has also been reported that a creatine plus protein supplement had no significant effect on improving isokinetic muscle function in middle-aged and older men *(62)*. These results seem to indicate that the addition of protein and/or carbohydrate to creatine may not be superior to using monohydrate alone.

Many questions are yet unanswered on the various formulations of creatine and their usefulness. At this point the exact mechanisms by which maximal exercise performance is enhanced by creatine are not fully understood, but it is likely that the insulin response is not the sole regulating mechanism. Previous research runs a vast spectrum, as it 1) has reported that creatine uptake is regulated by sodium and not insulin *(63)*, and 2) that the insulin response is important for creatine accumulation in the muscle, and that adding carbohydrate and protein best potentiates this response *(56)*. It is possible that multiple transporter mechanisms are involved in creatine uptake, and further research on these mechanisms is needed to best delineate if any other forms of creatine might prove more effective at promoting strength and lean body mass than monohydrate. The timing of ingestion of various combinations may also play a key role in their effectiveness of the creatine combination on strength and lean body mass. An investigation of protein and carbohydrate creatine combinations and various ingestion times seems pertinent. In addition, it could be that there is a threshold of muscle creatine retention that is reached, over which no significant changes in strength and lean body mass occur, that is traditionally obtained by creatine monohydrate supplementation alone.

One of the more popular ergogenic aids used in combination with creatine, as well as in isolation, is β-hydroxy-β-methylbutyrate (HMB). HMB is a metabolite of the amino acid leucine that has been shown to increase lean body mass and strength when combined with resistance training *(64,65)* as well as reported to produce no adverse health effects when combined with creatine *(66)*. When used in combination with creatine, two studies have shown differing effects. Investigators reported that the HMB/creatine combination produced greater increases in lean body mass and strength than either supplement did alone *(67)*. This group theorized that HMB and creatine are able to work synergistically because they work by different mechanisms; i.e., HMB works by slowing muscle-protein breakdown, whereas creatine increases lean body mass by increasing cellular water content. However, other investigators reported no significant difference from the control group in elevating aerobic or anaerobic capacity in an HMB/creatine supplementation group following 6 wk of taking the supplement *(68)*.

Recently, interest has developed in the use of β-alanine in conjunction with creatine. β-alanine is a derivative of the cytoplasmic dipeptide carnosine. Carnosine plays a major role in pH buffering in muscle, and is found in greater amounts in type II muscle fibers. Harris and associates reported a 64% increase in carnosine levels with 15 d of β-alanine. Of the two precursors to carnosine, β-alanine and free histadine, β-alanine seems to be the limiting factor in carnosine synthesis *(69)*. Therefore, it has been theorized that supplementing creatine with β-alanine will have an improved effect on delaying muscle fatigue owing to decreases in pH. Although unpublished in manuscript form to date, two recent abstract presentations have investigated the effects of creatine and β-alanine. Hill and associates reported significantly greater cycling work completed with creatine alone, β-alanine alone, and in combination than placebo, but no additive effects were noted *(70)*. Hoffman and colleagues reported significant differences from placebo in creatine and creatine plus β-alanine groups *(71)*. However, creatine plus β-alanine did produce significantly greater changes in both lean mass and percentage of body fat than placebo or creatine alone.

Another recent supplement added to creatine is that of cinnamon. As discussed earlier, this formulation approach has been marketed and combined with creatine ethyl ester. Claims have been made that an aqueous form of cinnamon called CinnulinPF® (concentrated water-soluble cinnamon extract that stimulates glucose uptake and glycogen synthase) that is added to numerous supplement products will help to drive creatine into the muscle. No research to date has been conducted on the safety and efficacy of cinnamon in improving muscle retention of creatine. Cinnamon has been shown to improve glucose uptake by enhancing the insulin signaling pathway *(72,73)*. Although these findings have been initially applied to insulin-resistant diabetics, no research has supported the claims that creatine retention is enhanced by using the product. Furthermore, the combination of creatine with cinnulin may exacerbate (*note*: what does the author mean by exacerbate muscle creatine uptake??) muscle creatine uptake by stimulating skeletal-muscle-mediated glucose uptake and concomitant upregulation in the creatine transporter.

Again, to date, although there are several published studies on the effects of cinnulin extract, there are no training studies combining cinnulin with creatine in attempt to increase muscle creatine. In fact, even though the scientific jury is still out, various creatine research leaders are skeptical about the postulated transporter mechanisms of cinnulin combined with creatine ethyl ester, which openly refutes the current scientific knowledge regarding the creatine transport mechanistic process. Creatine ethyl ester is a new formulation that combines creatine monohydrate with an ester. Esters are organic compounds that are formed by esterification, the reaction of carboxylic acid and alcohols. Creatine ethyl ester is supposedly a membrane permeable form of creatine that theoretically can enter the cells without having to use the creatine transporter molecules. There seems to be a limit for muscle creatine uptake when the transport system is downregulated below a concentration of 150 mmol/L. Whether the muscle creatine uptake resulting from creatine ethyl ester is any higher than those achievable with creatine monohydrate based on a 30-d period of supplementation (at 5 g/d) is virtually unknown owing to the absence of published research.

Typically, when taken alone, creatine has been administered in the form of creatine monohydrate dissolved in some form of a liquid. Other forms have been investigated including serums and effervescents. Effervescent forms of creatine, including dicreatine citrate, have been investigated because of better solubility and enhanced palatability in comparison with creatine monohydrate *(74)*. Investigators have concluded that the effervescent form of creatine used showed no significant difference to creatine monohydrate in terms of muscle retention, but was not as effective as a creatine + dextrose group *(55)*. In a study with similar findings, investigators reported that an effervescent creatine, ribose, and glutamine supplement did not enhance muscular strength or endurance when compared with placebo over the course of 8 wk of training *(75)*. Two investigations on serum creatine formulations have reported that serum is not effective in increasing muscle-free creatine *(39)*, and that within the serum, 98% of the active creatine was converted to creatinine *(76)*.

Many other miscellaneous forms of creatine have been tried that have not proved better than creatine monohydrate, including Mg^{2+}creatine chelate *(77)*, creatine phosphate *(23)*, creatine and glutamine *(78)*, and creatine combined with sodium bicarbonate *(79)*. In the former study, 6 d of creatine loading followed by sodium bicarbonate supplementation on the seventh day was found to be more effective at improving swim times than placebo. However, no comparison was made with monohydrate alone. Other miscellaneous additives have been tried, but none of these combinations have proved more effective than traditional creatine monohydrate in promoting gains in lbm or strength. In addition, α-lipoic acid, which enhances the function of insulin and therefore should enhance the amino acid deposition within the muscle, has been evaluated in combination with creatine in pigs. Muscle quality was examined and was found to be superior in the creatine monohydrate group than the creatine-α-lipoic acid group *(80)*.

The proposed mechanisms for each of these combinations seem to be logical, but as of yet have not been validated as effective. Glutamine is an amino acid that has been reported to enhance protein synthesis in humans *(81)*. It also appears that the nutrient may also increase cell volume and osmotic pressure *(63)*. Investigators have also hypothesized that: "Glutamine may promote cellular volume and osmolarity as a result of an insulin-dependent, sodium-dependent transport system *(77)*." Therefore creatine and glutamine might result in increased cellular swelling leading to enhanced muscle mass. However, their investigation revealed no significant differences between creatine and creatine + glutamine groups.

Mg^{2+}creatine chelate has been proposed as an alternative method of getting creatine into the cell by making it a chelate containing a cation. It was thought that the creatine would then be able to enter the cell through a ligand-gated channel that only allows cations to travel through. However, this method proved no better at improving performance than creatine monohydrate in the Selsby investigation *(75)*.

Bicarbonate is one of the primary hydrogen ion buffers in the body that helps to prevent change in pH within a cell. Because a change in pH is one of the leading contributors to muscle fatigue, it is rationale to hypothesize that bicarbonate supplementation could increase work performance. McNaughton and associates reported that sodium bicarbonate increased maximal work and power output capabilities in women *(82)*. It is possible that if the sodium transport mechanism is in fact highly important to creatine uptake, that adding sodium bicarbonate could be beneficial and allow the two to work synergistically. However, another investigation reported no significant advantage to this supplementation over monohydrate *(78)*.

Phosphate is important in the body in combination with creatine; these two join to form phosphocreatine, the storage form of high-energy phosphate within the body. Phosphocreatine is used within skeletal muscle to generate adenosine triphosphate to provide energy for the cells. Creatine phosphate is formed by adding phosphate salts to creatine through the chelation process. While researchers have shown creatine phosphate to be as effective as creatine monohydrate at increasing strength and lean body mass, if it is more difficult and more expensive to produce which favors creatine monohydrate as a economical but still an effective ergogenic option *(76)*. In addition, other investigators have combined glucose, sodium and taurine in an effort to enhance creatine uptake in the muscle cell *(83)*.

It appears that an optimal nutritional supplement aimed at promoting gains in strength and lean body mass, such as a quality creatine combination, should have several distinct properties. First, it needs to be absorbed intact through the intestine into the blood optimizing creatine transport properties, including chemical and enzymatic stability, solubility, and low clearance by the liver or kidney, permeation across biological membranes, potency, and safety *(84)*. Second, whereas more research is warranted, it should exhibit properties that enhance muscle function such as: buffering hydrogen ions and preventing pH decline, increasing cell volume, acting on the proper transport mechanisms for optimal uptake and be sodium dependent mediated by countless insulin options *(85–90)*. Consequently, it is recommended that creatine ingestion occur with a high carbohydrate drink and/or with a combined carbohydrate/protein supplement in order to elevate insulin and promote creatine uptake *(39,40,56,57,91)*. Finally, although not related to efficacy, in practical terms creatine combination practices should be consumer friendly with additional attributes such as palatability and affordability.

2.3. Foundational Creatine Ergogenic Efficacy

As creatine is one of the most popular nutritional supplements to appear in the consumer market, a plethora of research has been conducted to substantiate the ergogenic efficacy of this product. Although other professionals in this text will cover this premise in detail, only a brief general scenario will be provided herein regarding creatine's ergogenic efficacy to date. Specifically, there are approx 700 peer-viewed manuscripts that have investigated the effects of creatine supplementation on a variety of exercise and/or training activities with significant ergogenic outcomes *(92)*. In the majority of these studies to date, there is an overall increase of 10–15% related to a variety of performance outcomes. The following factors have been reported to improved specific performance outcomes related to short-term creatine supplementation: (1) maximal power/strength ranging from increases of 5–15%, (2) work performed during sets of maximal effort muscle contractions at increases of 5–15%, (3) single-effort sprint performance increases from 1 to 5%, and (4) increased work performed during repetitive sprint performance ranging from 5 to 15% *(1,92)*. When evaluating the ergogenic outcomes of long-term creatine ingestion the findings are comparable with that of short-term ingestion. Typically, the overall enhancement of training has been reported to promote improved gains in strength and performance from 5 to 15% *(1,92)*. When evaluating research related to body mass, the peer reviewed findings are consistent indicating a 1–2 kg improvement within the first 7 d of creatine supplementation in the majority of studies. Furthermore, individuals ingesting creatine during long-term training studies normally increase body mass and/or lean muscle mass up to twice as much as individuals ingesting placebos over a 4–12-wk period. Ironically, no training study to date has demonstrated decrements in exercise capacity because of the ingestion of creatine supplementation. Although a vast number of studies support these outcomes it should also be noted that some individuals respond better to creatine supplementation than a group often referred to as nonresponders *(93,94)*. Although a small majority of investigators report no significant ergogenic outcomes, there is overwhelming scientific based evidence promoting creatine as a safe and effective nutritional supplement for a multitude of training activities and populations *(1,92)*.

2.4. Future Creatine Research Options

Without question the quality and amount of scientific based creatine research has established a vital foundation regarding a very popular but often misunderstood nutritional supplement. However, although many viable questions have been answered to date, the research regarding the effects of creatine supplementation is far from over and in some instances just beginning. Specifically, one of the most extensively studied topics has focused on short-term ergogenic performance benefits of creatine ingestion. However, in a closely related area, limited research has been conducted relative to cycling on and off creatine to evaluate the ergogenic performance effects regarding long-term creatine ingestion. Additionally, time- and labor-intensive research involving long-term creatine ingestion and its effects on markers of overtraining and injury outcomes is limited and warrants extensive investigation. Although various anaerobic athletic/exercise populations have been examined regarding creatine supplementation, some experts contend there is ergogenic value investigating endurance athletes (i.e., runners and cyclists) involved in intense interval training bouts. The contention herein is that athletes can improve their anaerobic threshold and thus aerobic capacity by incorporating creatine supplementation with intense interval training thereby elevating stores of the initial energy system, phosphocreatine. In addition, creatine ingestion with carbohydrate loading helps maximize glycogen retention and therefore optimizing interval training and aerobic performance outcomes. A major challenge associated with conducting creatine research with collegiate athletes surfaced when the NCAA (August 1, 2000) mandated that creatine products could no longer be disseminated to athletes by any school representative (strength coaches, athletic trainers, athletic coaches, researchers, and so on). Specifically, only nonmuscle-building nutritional supplements can be provided to student-athletes for the sole purpose of providing additional calories and/or electrolytes, provided the supplements are not classified as NCAA banned substances. Nonmuscle-building nutritional supplements were categorized as: (1) vitamins and minerals, (2) energy bars, (3) calorie-replacement drinks, and (4) electrolyte-replacement drinks. A troubling aspect of this mandate was that it did not revolve around safety related concerns but in fact equality in competition as some athletic conferences/institutions could not afford to supply their athletes with creatine and/or other muscle-building products. Another unfortunate outcome of this stipulation was the monitoring limitations placed on qualified professionals regarding the athletes' consumption of creatine products in this environment. Who would be better suited to educate and monitor these athletes than the qualified professionals they work closely with on a daily basis? Whereas athletes can legally purchase and ingest creatine on their own, this policy has hindered investigators focused on conducting much needed quality controlled research by monitoring pertinent extraneous variables (i.e., creatine quality, dosage protocols, timing of ingestion, and formulations). This NCAA mandate has stifled the amount and quality of long-term safety/performance creatine research with collegiate athletes that demands time, labor, and financial intensive commitment.

The potential therapeutic values of creatine have elevated its safety and efficacy primarily because of the medical profession utilizing it with various clinical and/or special populations previously noted. However, many creatine experts believe that research in this area has not come close to the tip of the iceberg. Specifically, much of the work with neuromuscular diseases and muscle wasting should extend the focus to varying stages of specific long-term afflictions (i.e., multiple sclerosis, muscular dystrophy, amyotrophic lateral sclerosis, and HIV) in effort to slow the debilitating process of these conditions thereby enhancing a person's functional quality of life. Another area of creatine research that is limited includes muscle tissue recovery after surgery and/or soft tissue injuries from various traumatic events. It is possible that some of the more common muscular injuries (muscle strains/sprains, and so on) that occur in athletic and/or exercise environments would benefit with faster recoveries owing to viable creatine supplementation ingestion strategies. Another very interesting approach could consider using creatine in weight loss studies to reduce body fat while simultaneously attempting to enhance and/or maintain lean muscle mass. A very unique approach to creatine ingestion, as research designs typically promote this supplement as an effective weight gaining option for select populations, related to increases in total body weight not solely body composition reductions as a viable weight loss technique.

Two final future interdependent creatine research suggestions include the elusive journey to discover the most viable supplement ingredients to promote muscle-creatine uptake and the biochemical mechanisms that enhance this process. Although creatine combinations have already been addressed in this chapter, investigators continue to search for rational scientific explanations and mechanisms to improve the upregulation of creatine levels even though some experts believe saturation levels are limited thereby dictating everyone's uptake capabilities. Even with the existing creatine uptake research to date, investigators should attempt to explore safe and effective scientific methods to progressively increase creatine saturation levels to not only enhance athletic/exercise performance outcomes but more importantly a vast array of clinical based health conditions. Dr. Darryn Willoughby, Professor of Exercise Biochemistry and Molecular Biology and a leading expert regarding the mechanistic aspects of creatine supplementation provides the following details related to existing and future research in this challenging area:

The quest for different formulations of creatine that will enhance muscle creatine uptake and provide performance improvements are ongoing; however, most of these new formulations have little or no scientific data to support the claims made by most of the companies who sell these products. In this light, what cannot be overlooked is that whole body creatine retention is dependent primarily on rates of creatine uptake and intramuscular creatine content, and to a lesser extent, the slow degradation of creatine into creatinine. Creatine uptake into the muscle is dependent on the creatine transporter, a membrane-spanning protein that transfers creatine from the blood into the muscle fibers. The regulation of the creatine transporter protein appears to be of paramount importance in controlling intramuscular creatine levels. It seems that creatine uptake is actually inhibited with prolonged exposure to high plasma creatine levels, which may be owing to decreased activity of creatine transporter. This suggests that creatine uptake is actually dependent on intracellular creatine concentrations, and not extracellular creatine concentrations. Elevated plasma creatine levels promote increases in muscle creatine uptake. However, there may be a point in which the resultant intramuscular creatine concentrations may play a role in inhibiting creatine transport regardless of plasma creatine levels.

The other aspect to consider with various creatine formulations is a way to exacerbate the mechanisms in which longer-term creatine supplementation stimulates skeletal muscle hypertrophy. Creatine enhances short-term, anaerobic endurance through its inherent ability to enhance muscle bioenergetics. In addition, creatine indirectly promotes muscle anabolism with longer-term supplementation coupled with resistance training by extending exercise output, again through enhanced muscle bioenergetics. As a result, muscles then compensate for the increased mechanical load through the production of new muscle proteins. These newly added proteins promote hypertrophy, thereby allowing muscles to generate greater amounts of force and power. Long-term (10–15 wk) studies have been done showing 5–6 g/d of creatine combined with heavy resistance training is preferentially more effective than placebo/control at increased muscle strength, muscle mass, and types I, IIa, and IIab muscle fibers *(95)*, Types I and IIa myosin heavy chain isoform mRNA and protein expression, myofibrillar protein content *(44)*, and satellite cell number and muscle myonuclei content *(96)*. Additionally, it has been shown that creatine preferentially increased the mRNA and protein expression of the myogenic regulatory factors (MRF4) and myogenin, and DNA-binding proteins that serve as transcription factors in upregulating the expression of muscle-specific genes. Additionally, myogenin and MRF4 expression was directly correlated to the mRNA expression of the skeletal muscle isozyme of creatine kinase *(97)*. These studies have been the first to actually demonstrate a direct hypertrophic link to creatine supplementation through molecular mechanisms linked to muscle hypertrophy through both muscle-specific gene and protein expression and increases in satellite cell activation.

2.5. Conclusion

Creatine products have become one of the most popular nutritional supplements available today despite various anecdotal misconceptions that misrepresent the scientific safety and efficacy data available in the current research. Quality creatine supplementation research outcomes have supported aspects such endogenous safety, augmented muscle-creatine uptake, enhanced short/long-term exercise and training performance outcomes, as well as therapeutic benefits associated with various clinical populations. Although foundational creatine research options have advanced over the past 15 yr, the stage has been set to continue replicating/supporting these specific areas in hope of igniting new investigative directions. Persistent researchers should strive to further maximize safe strategies to upregulate muscle creatine, which could have monumental positive clinical and ergogenic implications. Key creatine factors that will help accomplish this challenge include but are not limited to the following aspects: (1) quality, purity, formulation, (2) innovative/precise dosage strategies, and (3) combination ingredients. Although creatine products have lasted the scientific test of time to date, the evolving research cycle seems promising as well as exciting regarding this very safe and effective nutritional supplement.

2.6. Common Creatine Practical Applications

1. Creatine and other nutritional supplements are not a complete substitute for a well balanced nutrient dense diet. However, nutritional supplementation strategies, in addition to a nutrient dense diet, are vital in assisting the athlete in replacing the necessary caloric requirements lost through high-intensity energy expenditure. A valuable rule to implement is that a supplement is indeed a supplement to enhance caloric intake, not a complete replacement for a nutrient dense diet.
2. In addition to a quality nutrient dense diet combined with creatine ingestion, athletes/exercisers should take a multivitamin daily (with iron for female athletes).
3. To load or not to load? If the goal is to saturate muscle-creatine levels rapidly, the loading cycle is effective in accomplishing this goal (i.e., 20–25 g/d every 4 h). However, if creatine muscle uptake is not urgent, ingesting creatine (5 g/d) over the course of 28–30 d can accomplish similar outcomes.

4. Research indicates that creatine loading combined with carbohydrate/protein loading helps maximize glycogen retention and thereby successful creatine muscle saturation.
5. Nutritional timing is imperative for any dietary strategy. Therefore, after proper nutrient dense caloric intake has occurred, it is recommended to ingest creatine 1 h before training/competition (saturate uptake levels) and within 1 h after the exercise bout to replenish creation stores. In addition to creatine supplementation, maintaining energy balance to reduce catabolic states, athletes are encouraged to consume 50–100 g of carbohydrate and 30–40 g of protein 30–60 min before exercise. Furthermore, because some individuals lose their appetites after intense training, a carbohydrate/protein supplement snack is recommended within 30 min after the exercise bout followed by a nutrient dense meal within the critical 2 h caloric recovery window.
6. Because it is difficult to consume large quantities of food in one setting and difficult to maintain quality energy balance, athletes are encouraged to eat four to six meals per day with a quality creatine supplement. Ingesting creatine carbohydrate/protein snacks between meals can help offset extreme energy expenditure levels derived from intense duration training.
7. The most relevant and beneficial ergogenic effects from creatine supplementation are derived from intermittent activities that require short intense—powerful—explosive—repetitive movements that utilize the phosphocreatine energy system. Therefore, any training and/or competitive exercise performance bouts demanding these factors will benefit from creatine ingestion (i.e., 60–200 m running/cycling sprints, football, basketball, volleyball, soccer, swimming, and baseball). A significant body of research indicates that creatine supplementation during training helps increase high-intensity intermittent work output, and promote greater gains in strength and muscle mass.
8. Endurance athletes implementing interval workouts in their training and interested in maximizing glycogen availability before competitive events can derive indirect benefits from creatine supplementation.
9. Available research indicates that short/long-term creatine supplementation is safe and helps lessen the incidence of injury in athletes whereas more research is needed to investigate creatine to enhance recovery from injuries.
10. Creatine supplementation has been shown to provide therapeutic benefits for a number of clinical populations but additional research is warranted. When working with clinical populations it is highly recommended to consult ones' physician before implementing any creatine supplemental strategies.

11. Although gains in body mass and lean muscle mass do not occur as rapidly as men, recent well-controlled research has shown that women attain short and long-term ergogenic benefits from creatine supplementation *(98)*. Therefore, women ingesting creatine products can accomplish strength and muscle mass benefits from training over a greater duration of time.
12. According to current research, children and young adolescents (teenagers) can derive ergogenic benefits from creatine supplementation whereas also providing a safe alternative to experimenting with anabolic steroids. Although no research to date indicates that creatine ingestion is harmful to children or adolescents, there are less scientific based studies conducted with these populations. Consequently, creatine researchers suggest that parents and adolescents consider this supplement method only if the following stipulations are met first: (a) parental permission is given after potential benefits and side effects are presented and comprehended, (b) after puberty and when participating in intense training, (c) supplementation ingestion is supervised/monitored by qualified professionals (athletic trainers, strength training specialists, physicians, athletic coaches, and so on), (d) quality creatine is used based on recommended dosages, and (e) in conjunction with well-balanced nutrient dense diet *(99)*.
13. Various creatine formulations combined with other nutritional supplement ingredients do not meet NCAA requirements. Therefore, athletes should consult a nutrition specialist, certified athletic trainer, certified strength and conditioning specialist, and/or other qualified professionals to ensure the creatine-based combination supplement(s) are not banned by the NCAA.
14. Anecdotally, creatine ingestion has been unjustifiably linked to problems associated with various aspects of heat illness (i.e., dehydration, heat illness, and muscle cramping). Although current research has not supported these contentions, the truth of the matter is many individuals are in dehydrated state because of inadequate fluid consumption. Whether one takes creatine or not, ingesting at least 64 ounces of water/d (more during intense training in hot/humid climates) helps reduce dehydration concerns. The following hydration strategies are recommended: four to six cups of water/glucose electrolyte solution (GES) before training begins, six to eight cups of water/GES every 5–15 min during training, and replenish lost fluid amounts after training. The athlete's weight should be monitored closely throughout the training period in effort to reduce heat illness. The hydration standard is to drink three cups of water/GES for every pound lost.

ACKNOWLEDGMENTS

I would like to acknowledge the International Society of Sport Nutrition (ISSN) board of directors and advisor board members for their professional commitment in promoting Sport Nutrition and for developing the ISSN. My appreciation and thanks also goes out to the faculty members, students and subjects who have graciously contributed to the profession by supporting my research endeavors. Currently I am not affiliated with any nutritional supplement companies as a consultant but have attained various grants to conduct sport/exercise nutritional supplement research as a primary investigator and/or coinvestigator from the following organizations: Wellness Enterprises, iSatoria Global Technology, MetRx, Muscle Tech, Mannatech, Meta Response Sciences, Champion Nutrition, Cytodyne Technology, Natrol, SKW Trostberg, Weider, and the National Honey Board.

REFERENCES

1. Kreider RB, Leutholtz BC, Greenwood M. Creatine, in Nutritional Ergogenic Aids. Wolinsky I. and Driskell J. (eds.), CRC Press LLC: Boca Raton, FL, 2004, pp. 81–104.
2. Williams MH, Kreider R, Branch JD. Creatine: The power supplement. Champaign, IL: Human Kinetics Publishers, 1999, pp. 204–205.
3. Kreider R, Rasmussen C, Melton M, et al. Long-term creatine supplementation does not adversely affect clinical markers of health. Med Sci Sports Exerc 2000; 32(5):S134.
4. Vandenberghe K, Goris M, Van Hecke P, Van Leemputte M, Vangerven L, Hespel P. Long-term creatine intake is beneficial to muscle performance during resistance training. J Appl Physiol 1997; 83(6):2055–2063.
5. Tarnopolsky MA, Parise G, Fu MH, Parshad A, Speer O, Wallimann T. Acute and moderate-term creatine monohydrate supplementation does not affect creatine transporter mRNA or protein content in either young or elderly humans. Mol Cell Biochem 2003; 244(1–2):159–166.
6. Robinson TM, Sewell DA, Casey A, Steenge G, Greenhaff PL. Dietary creatine supplementation does not affect some haematological indices, or indices of muscle damage and hepatic and renal function. Br J Sports Med 2000; 34(4):284–288.
7. Poortmans JR, Auquier H, Renaut V, Durussel A, Saugy M, Brisson GR. Effect of short-term creatine supplementation on renal responses in men. Eur J Appl Physiol Occup Physiol 1997; 76(6):566–567.
8. Schilling BK, Stone MH, Utter A, et al. Creatine supplementation and health variables: a retrospective study. Med Sci Sports Exerc 2001; 33(2):183–188.
9. Taes YE, Delanghe JR, De Vriese AS, Rombaut R, Van Camp J, Lameire NH. Creatine supplementation decreases homocysteine in an animal model of uremia. Kidney Int 2003; 64:1331–1337.
10. Poortmans JR, Francaux M. Adverse effects of creatine supplementation: fact or fiction? Sports Med 2000; 30(3):155–170.
11. Kuehl K, Goldberg L, Elliot D. Effects of oral creatine monohydrate supplementation on renal function in adults. Med Sci Sports Exerc 2000; 32(5):S168.

12. Poortmans JR, Francaux M. Long-term oral creatine supplementation does not impair renal function in healthy athletes. Med Sci Sports Exerc 1999; 31(8):1108–1110.
13. Poortmans JR, Francaux M. Renal dysfunction accompanying oral creatine supplements. Lancet 1998; 352(9123):234.
14. Earnest CP, Almada A, Mitchell TL. Influence of chronic creatine supplementation on hepatorenal function. FASEB J 1996; 10:A790.
15. Volek JS, Mazzetti SA, Farquhar WB, Barnes BR, Gomez AL, Kraemer WJ. Physiological responses to short-term exercise in the heat after creatine loading. Med Sci Sports Exerc 2001; 33(7):1101–1108.
16. Kreider RB, Melton C, Rasmussen CJ, et al. Long-term creatine supplementation does not significantly affect clinical markers of health in athletes. Mol Cell Biochem 2003; 244(1–2):95–104.
17. Greenwood M, Farris J, Kreider R, Greenwood L, Byars A. Creatine supplementation patterns and perceived effects in select division I collegiate athletes. Clin J Sport Med 2000; 10(3):191–194.
18. Oopik V, Paasuke M, Timpmann S, Medijarnen L, Erelinen J, Smirnova T. Effect of creatine supplementation during rapid body mass reduction on metabolism and isokinetic muscle performance capacity. Eur J Appl Physiol Occup Physiol 1998; 78(1):83–92.
19. Kreider RB, Ferreira M, Wilson M, et al. Effects of creatine supplementation on body composition, strength, and sprint performance. Med Sci Sports Exerc 1998; 30(1):73–82.
20. Volek JS, Mazzetti SA, Farquhar WB, Barnes BR, Gomez AL, Kraemer WJ. Physiological responses to short-term exercise in the heat after creatine loading. Med Sci Sports Exerc 2001; 33(7):1101–1108.
21. Zehnder M, Rico-Sanz J, Kuhne G, Dambach M, Buchi R, Boutellier U. Muscle phosphocreatine and glycogen concentration in humans after creatine and glucose polymer supplementation measured noninvasively by 31P and 13C-MRS. Med Sci Sports Exerc 1998; 30(5):S264.
22. Mihic S, MacDonald JR, McKenzie S, Tarnopolsky. Acute creatine loading increases fat-free mass, but does not affect blood pressure, plasma creatinine, or CK activity in men and women. Med Sci Sports Exerc 2000; 32(2):291–296.
23. Peeters BM, Lantz CD, Mayhew JL. Effect of oral creatine monohydrate and creatine phosphate supplementation on maximal strength indices, body composition, and blood pressure. J Strength Cond Res 1999; 13(1):3–9.
24. Stone MH, Schilling BK, Fry AC, et al. A retrospective study of long-term creatine supplementation on blood markers of health. J Strength Cond Res 1999; 13(4):434.
25. Schroeder C, Potteiger J, Randall J, et al. The effects of creatine dietary supplementation on anterior compartment pressure in the lower leg during rest and following exercise. Clin J Sport Med 2001; 11(2):87–95.
26. Potteiger JA, Carper MJ, Randall JC, Magee LJ, Jacobsen DJ, Hulver MW. Changes in lower leg anterior compartment pressure before, during, and after creatine supplementation. J Athl Train 2002; 37:157–163.
27. Hile AM, Anderson JM, Fiala KA, Stevenson JH, Casa DJ, Maresh CM. Creatine supplementation and anterior compartment pressure during exercise in the heat in dehydrated men. J Athl Train 2006; 41(1):30–35.
28. Watson G, Casa DJ, Fiala KA, et al. Creatine use and exercise heat tolerance in dehydrated men. J Athl Train 2006; 41(1):18–29.

29. Greenwood M, Kreider RB, Greenwood L, Byars A. Cramping and injury incidence in collegiate football players are reduced by creatine supplementation. J Athl Train 2003; 38(3):216–219.
30. Greenwood M, Kreider R, Greenwood L, Willoughby D, Byars A. Creatine supplementation does not increase the incidence of injury or cramping in college baseball players. J Exerc Physiol online 2003; 6(4):16–22.
31. Kilduff LP, Georgiades N, James RH, et al. The effects of creatine supplementation on cardiovascular, metabolic, and thermoregulatory responses during exercise in the heat in endurance-trained humans. Int J Sport Nutr Exerc Metab 2004; 14(4): 443–460.
32. Yoshizumi WM, Tsourounis C. Effects of creatine supplementation on renal function. J Herb Pharmcother 2004; 4(1):1–7.
33. Greenwood M, Kreider RB, Melton C, et al. Creatine supplementation during college football training does not increase the incidence of cramping or injury. Mol Cell Biochem 2003; 244(1–2):83–88.
34. LaBotz M, Smith BW. Creatine supplement use in an NCAA Division I athletic program. Clin J Sport Med 1999; 9(3):167–169.
35. Sipila I, Rapola J, Simell O, Vannas A. Supplementary creatine as a treatment fo gyrate atrophy of the choroid and retina. N Engl J Med 1981; 304:867–870.
36. Vannas-Sulonen K, Sipila I, Vannas A, Simell O, Rapola J. Gyrate atrophy of the choroid and retina. A five-year follow-up of creatine supplementation. Ophthalmology 1985; 92(12):1719–1727.
37. Benzi G. Is there a rationale for the use of creatine either as nutritional supplementation or drug administration in humans participating in a sport? Pharmacol Res 2000; 41:255–264.
38. Persky AM, Brazeau GA. Clinical pharmacology of the dietary supplement creatine monohydrate. Pharmacol Rev 2001; 53:161–176.
39. Kreider RB, Willoughby DS, Greenwood M, Parise G, Payne E, Tarnopolsky MA. Effects of serum creatine supplementation on muscle creatine content. J Exerc Physiol online 2003; 6:24–33.
40. Greenwood M, Kreider RB, Rasmussen C, Almada AL, Earnest CP. D-Pinitol augments whole body creatine retention in man. J Exerc Physiol Online 2001; 4:41–47.
41. Gill ND, Hall RD, Blazevich AJ. Creatine serum is not as effective as creatine powder for improving cycle sprint performance in competitive male team-sport athletes. J Strength Cond Res 2004; 18:272–275.
42. Michaelis J, Vukovich MD. Effect of two different forms of creatine supplementation on muscular strength and power. Med Sci Sports Exerc 1998; 30:S272.
43. Hultman E, Soderlund K, Timmons JA, Cederblad G, Greenhaff PL. Muscle creatine loading in men. J Appl Physiol 1996; 81:232–237.
44. Willoughby DS, Rosene J. Effects of oral creatine and resistance training on myosin heavy chain expression. Med Sci Sports Exerc 2001; 33:1674–1681.
45. Burke DG, Chilibeck PD, Yu PH, Candow DG. (In Press) Development of an optimal creatine dose during loading based on lean body mass. Appl Physiol Nutr Metab.
46. Borsheim E, Aarsland A, Wolfe RR. Effect of an amino acid, protein, and carbohydrate mixture on net muscle protein balance after resistance exercise. Int J Sport Nutr Exerc Metab 2004; 14:255–271.

47. Borsheim E, Cree MG, Tipton KD, Elliott TA, Aarsland A, Wolfe RR. Effect of carbohydrate intake on net muscle protein synthesis during recovery from resistance exercise. J Appl Physiol 2004; 96:674–678.
48. Miller SL, Tipton KD, Chinkes DL, Wolf SE, Wolfe RR. Independent and combined effects of amino acids and glucose after resistance exercise. Med Sci Sports Exerc 2003; 35:449–455.
49. Bohe J, Low A, Wolfe RR, Rennie MJ. Human muscle protein synthesis is modulated by extracellular, not intramuscular amino acid availability: a dose-response study. J Physiol 2003; 552:315–324.
50. Wolfe RR. Regulation of muscle protein by amino acids. J Nutr 2002; 132: 3219S–3224S.
51. Volek JS, Ratamess NA, Rubin MR, et al. The effects of creatine supplementation on muscular performance and body composition responses to short-term resistance training overreaching. Eur J Appl Physiol 2004; 91:628–637.
52. Balsom PD, Soderlund K, Ekblom B. Creatine in humans with special reference to creatine supplementation. Sports Med 1994; 18:268–280.
53. Havenetidis K, Matsouka O, Cooke CB, Theodororis A. The use of varying creatine regimes on sprint cycling. J Sports Sci Med 2003; 2:88–97.
54. Burke DG, Chilibeck PD, Parise G, Candow DG, Mahoney D, Tarnopolsky M. Effect of creatine and weight training on muscle creatine and performance in vegetarians. Med Sci Sports Exerc 2003; 35:1946–1955.
55. Greenwood M, Kreider R, Earnest C, Rasmussen C, Almada A. Differences in creatine retention among three nutritional formulations of oral creatine supplements. J Exerc Physio Online 2003; 6(2):37–43.
56. Steenge GR, Simpson EJ, Greenhaff PL. PRO- and CHO-induced augmentation of whole body creatine retention in humans. J Appl Physiol 2000; 89: 1165–1171.
57. Green AL, Hultman E. Carbohydrate ingestion augments skeletal muscle creatine accumulation during creatine. Am J Physiol Endocrinol Metab 1996; 34(5): E821–E826.
58. Burke DG, Chilibeck PD, Davison KS, Candow DG, Farthing J, Smith-Palmer T. The effect of whey PRO supplementation with and without creatine monohydrate combined with resistance training on lean tissue mass and muscle strength. Int J Sport Nutr Exerc Metab 2001; 11:349–364.
59. Tarnopolsky MA, Parise G, Yardley NJ, Ballantyne CS, Olatunji S, Phillips SM. Creatine-dextrose and PRO-dextrose induce similar strength gains during training. Med Sci Sport Exerc 2001; 33(12):2044–2052.
60. Chromiak JA, Smedley B, Carpenter W, et al. Effect of a 10-week strength training program and recovery drink on body composition, muscular strength and endurance, and anaerobic power and capacity. Nutrition 2004; 20:420–427.
61. Theodorou AS, Havenetidis K, Zanker CL, et al. Effects of acute creatine loading with or without CHO on repeated bouts of maximal swimming in high-performance swimmers. J Strength Cond Res 2005; 19(2):265–269.
62. Carter JM, Bemben DA, Knehans AW, Bemben MG, Witten MS. Does nutritional supplementation influence adaptability of muscle to resistance training in men aged 48 to 72 years? J Geriatric Phys Ther 2005; 28(2):40–47.

63. Willott CA, Young ME, Leighton B, et al. Creatine uptake in isolated soleus muscle: kinetics and dependence on sodium, but not on insulin. Acta Physiol Scand 1999; 166:99–104.
64. Kreider RB. Dietary supplements and the promotion of muscle growth with resistance exercise. Sports Med 1999; 27(2):97–110.
65. Nissen S, Sharp R, Ray M, et al. Effect of leucine metabolite β-hydroxy-β-methyl-butyrate on muscle metabolism during resistance-exercise training. J Appl Physiol 1996; 81(5):2095–2104.
66. Crowe MJ, O'Conner DM, Lukins JE. The effects of β-hydroxy-β-methylbutyrate (HMB) and HMB/creatine supplementation on indices of health in highly trained athletes. Int J Sport Nutr Exerc Metab 2003; 13:184–197.
67. Jowko E, Ostaszewski P, Jank M, et al. Creatine and β-hydroxy-β-methylbutyrate (HMB) additively increase lean body mass and muscle strength during a weight-training program. Nutrition 2001; 17:558–566.
68. O'Conner DM, Crowe MJ. Effects of β-hydroxy-β-methylbutyrate and creatine monohydrate supplementation on the aerobic and anaerobic capacity of highly trained athletes. J Sports Med Phys Fitness 2003; 43:64–68.
69. Harris RC, Tallon MJ, dunnett M, et al. The absorption of orally supplied β-alanine and its effect on muscle carnosine synthesis in human vastus lateralis. Amino Acids 2006; 30:279–289.
70. Hill CA, Chester A, Harris RC, et al. The effect of beta-alanine and creatine monohydrate supplementation on muscle composition and exercise performance. Med Sci Sports Exerc 2005; 37(5) Suppl:S348.
71. Hoffman JR, Ratamess NA, Kang J, Mangine G, Faigenbaum AD, Stout, JR. Effect of creatine and beta-alanine supplementation on performance and endocrine responses in strength/power athletes. Med Sci Sports Exerc 2006; 38(5) Suppl:S126.
72. Qin B, Nagasaki M, Ren M, Bajotto G, Oshida Y, Sato Y. Cinnamon extract (traditional herb) potentiates in vivo insulin-regulated glucose utilization via enhancing insulin signaling in rats. Diabetes Res Clin Pract 2003; 62:139–148.
73. Imparl-Radosevich J, Deas S, Polansky MM, et al. Regulation of PTP-1 and insulin receptor kinase by fractions from cinnamon: implications for cinnamon regulation of insulin signaling. Horm Res 1998; 50:177–182.
74. Ganguly S, Jayappa S, Dash AK. Evaluation of the stability of creatine in solution prepared from effervescent creatine formulations. AAPS Pharm Sci Tech 2003; 4(2):1–10.
75. Falk DJ, Heelan KA, Thyfault JP, Koch AJ. Effects of effervescent creatine, ribose, and glutamine supplementation on muscular strength, muscular endurance, and body composition. J Strength Cond Res 2003; 17(4):810–816.
76. Dash AK, Sawhney A. A simple LC method with UV detection for the analysis of creatine and creatinine and its application to several creatine formulations. J Pharm Biomed Anal 2002; 29:939–945.
77. Selsby JT, DiSilvestro RA, Devor ST. Mg^{2+}-creatine chelate and a low-dose creatine supplementation regimen improve exercise performance. J Strength Cond Res 2004; 18(2):311–315.
78. Lehmkuhl M, Malone M, Justice B, et al. The effects of 8 weeks of creatine monohydrate and glutamine supplementation on body composition and performance measures. J Strength Cond Res 2003; 17(3):425–438.

79. Mero AA, Keskinen KL, Malvela MT, Sallinen JM. Combined creatine and sodium bicarbonate supplementation enhances interval swimming. J Strength Cond Res 2004; 18(2):306–310.
80. Berg EP, Maddock KR, Linville ML. Creatine monohydrate supplemented in swine finishing diets and fresh pork quality: evalutating the cumulative effect of creatine monohydrate and alpha-lipoic acid. J Anim Sci 2003; 81:2469–2474.
81. Hankard RG, Haymond MW, Darmaun D. Effect of glutamine on leucine metabolism in humans. Am J Physiol 1994; 267:E343–E355.
82. McNaughton LR, Ford S, Newbold C. Effect of sodium bicarbonate ingestion on high intensity exercise in moderately trained women. J Strength Cond Res 1997; 11(2):98–102.
83. Krieder R, Ferreira M, Wilson M, et al. Effects of creatine supplementation on body composition, strength and sprint performance. Med Sci Sports Exerc 1997b; 30:73–82.
84. Liederer BM, Borchardt RT. Enzymes involved in the bioconversion of ester-based prodrugs. J Pharm Sci 2006; 95(6):1177–1195.
85. van Loon LJ, Murphy R, Oosterlaar AM, et al. Creatine supplementation increases glycogen storage but not GLUT-4 expression in human skeletal muscle. Clin Sci (Lond) 2004; 106:99–106.
86. Op ‘t Eijnde B, Richter EA, Henquin JC, Kiens B, Hespel P. Effect of creatine supplementation on creatine and glycogen content in rat skeletal muscle. Acta Physiol Scand 2001; 171:169–176.
87. Haughland RB, Chang DT. Insulin effects on creatine transport in skeletal muscle. Proc Soc Exp Biol Med 1975; 148:1–4.
88. Salomons GS, van Dooren SJ, Verhoeven NM, et al. X-linked creatine transporter defect: an overview. J Inherit Metab Dis 2003; 26:309–318.
89. Queiroz MS, Berkich DA, Shao Y, Lanoue K, Ismail-Beigi F. Thyroid hormone regulation of cardiac creatine content: role of Na+/creatine transporter expression. FASEB J 2001; 15:A477.
90. Young JC, Young RE, Young CJ. The effect of creatine supplementation on glucose transport in rat skeletal muscle. FASEB J 2000; 14:C249.
91. Green AL, Hultman E, Macdonald IA, Sewell DA, Greenhaff P. Carbohydrate feeding augments skeletal muscle creatine accumulation during creatine supplementation in humans. Am J Physiol 1996; 271:E821–E826.
92. Kreider RB. Effects of creatine supplementation on performance and training adaptations. Mol Cell Biochem 2003; 244:89–94.
93. Greenhaff PL, Bodin K, Soderlund K, Hultman E. Effect of oral creatine supplementation on skeletal muscle phosphocreatine resynthesis. Am J Physiol 1994; 266: E725–E730.
94. Syrotuik DS, Bell GJ. Acute creatine monohydrate supplementation: A descriptive physiological profile of responders versus non-responders. J. Strength Cond Res 2004; 18(3):610–617.
95. Volek J, Duncan N, Mazzetti S, et al. Performance and muscle fiber adaptations to creatine supplementation and heavy resistance training. Med Sci Sports Exerc 1999; 31(8):1147–1156.
96. Olsen S, Aagaard P, Kadi F, et al. Creatine supplementation augments the increase in satellite cell and myonuclei number in human skeletal muscle induced by strength training. J Physiol 2006; 573(Pt 2):525–534.

97. Willoughby D, Rosene J. Effects of oral creatine and resistance training on myogenic regulatory factor expression. Med Sci Sports Exerc 2003; 35(6):923–929.
98. Tarnopolsky MA, MacLennan DP. Creatine monohydrate supplementation enhances high-intensity exercise performance in males and females. Int J Sport Nutr Exerc Metab 2000; 10(4):452–463.
99. Kreider R. Sport Applications of Creatine, in Essentials of Sport Nutrition & Supplements. (Antonio J, Kalman D, Stout J, Greenwood M, Willoughby D. 1st ed.), Humana Press, Totowa, NJ, In Press 2007.

Index

D

E

F

G

H

I

K

L

M

N

Z

Kolibri 360 GmbH
Ferdinand-Jühlke-Straße 7
99095 Erfurt, Deutschland
produktsicherheit@kolibri360.de